SD329
Science: Level 3

The Open University

Signals and perception: the science of the senses

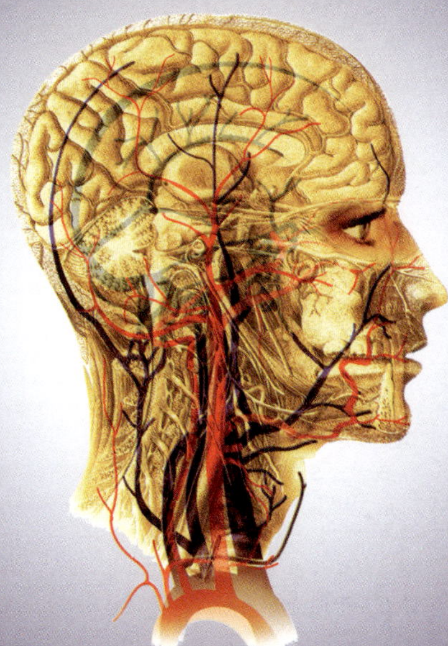

Block 3 Hearing and Balance

This publication forms part of the Open University course *SD329 Signals and Perception: the science of the senses*. The complete list of texts which make up this course can be found at the back. Details of this and other Open University courses can be obtained from the Student Registration and Enquiry Service, The Open University, PO Box 197, Milton Keynes
MK7 6BJ, United Kingdom (tel. +44 (0)845 300 60 90; email general-enquiries@open.ac.uk).

Alternatively, you may visit the Open University website at www.open.ac.uk where you can learn more about the wide range of courses and packs offered at all levels by The Open University.

To purchase a selection of Open University course materials visit www.ouw.co.uk, or contact Open University Worldwide, Walton Hall, Milton Keynes MK7 6AA, United Kingdom for a brochure (tel. +44 (0)1908 858793; fax +44 (0)1908 858787; email ouw-customer-services@open.ac.uk).

The Open University
Walton Hall, Milton Keynes
MK7 6AA

First published 2002. Second edition 2004. Third edition 2009.

Edited and designed by The Open University.

Typeset by The Open University.

Printed and bound in the United Kingdom by the Charlesworth Group, Wakefield.

The paper used in this publication is procured from forests independently certified to the level of Forest Stewardship Council (FSC) principles and criteria. Chain of custody certification allows the tracing of this paper back to specific forest-management units (see www.fsc.org).

ISBN 978 18487 3009 0

3.1

The SD329 Course Team

Course Team Chair
David Roberts

Course Manager
Yvonne Ashmore

Course Team Assistant
Margaret Careford

Authors
Mandy Dyson (Block 3)
Jim Iley (Block 6)
Heather McLannahan (Blocks 2, 4 and 5)
Michael Mortimer (Block 2)
Peter Naish (Blocks 4 and 7)
Elizabeth Parvin (Blocks 3 and 4)
David Roberts (Block 1)

Editors
Gilly Riley
Val Russell

Indexer
Jean Macqueen

OU Graphic Design
Roger Courthold
Jenny Nockles
Andrew Whitehead

CD-ROM and Website Production
Jane Bromley
Eleanor Crabb
Patrina Law
Kaye Mitchell
Brian Richardson
Gary Tucknott

Library
Judy Thomas

Picture Research
Lydia Eaton

External Course Assessors
Professor George Mather (University of Sussex)
Professor John Mellerio (University of Westminster)

Consultants
Michael Greville-Harris (Block 4, University of Birmingham)
Krish Singh (Block 2, Aston University)

BBC
Jenny Walker
Nicola Birtwhistle
Julie Laing
Jane Roberts

Reader Authors
Jonathan Ashmore (University College London)
David Baguley (Addenbrooke's Hospital, Cambridge)
Stanley Bolanowski (Syracuse University)
James Bowmaker (University College London)
Peter Cahusac (University of Stirling)
Christopher Darwin (University of Sussex)
Andrew Derrington (University of Nottingham)
Robert Fettiplace (University of Wisconsin)
David Furness (Keele University)
Michael Greville-Harris (University of Birmingham)
Carole Hackney (Keele University)
Debbie Hall (Institute of Hearing Research, Nottingham)
Anya Hurlbert (University of Newcastle upon Tyne)
Tim Jacob (University of Cardiff)
Tyler Lorig (Washington and Lee University)
Ian Lyon (Consultant)

Don McFerran (Essex County Hospital)
Keith Meek (University of Cardiff)
Tim Meese (Aston University)
Julian Millar (Queen Mary, University of London)
Peter Naish (Open University)
Robin Orchardson (University of Glasgow)
Alan Palmer (Institute of Hearing Research, Nottingham)
Krish Singh (Aston University)
Charles Spence (University of Oxford)
Rollin Stott (DERA Centre for Human Sciences)
Steve Van Toller (University of Warwick)
Stephen Westland (University of Derby)

BLOCK THREE

Hearing and Balance

Contents

1 Introduction

Humans are able to distinguish a remarkable range of sounds – from the complexity of a symphony to the simple ticking of a clock. Our sense of hearing provides us with a unique source of information about what is occurring in our immediate surroundings. Close your eyes for a moment and consider the sounds you hear and what they tell you about your environment. Our sense of hearing depends entirely on the sensory receptors of the inner ear known as hair cells. Similar hair cells are also responsible for our sense of equilibrium.

In humans, hearing commences when sound waves, generated by mechanical forces such as a bow being drawn across a string, or air being expelled across the vocal cords, cause the eardrum to vibrate. This vibration is passed, via three small bones in the middle ear, to the inner ear where the cochlea is located. It is here that sound energy is transduced into electrical signals by the hair cells. The hair cells send out a rapid-fire code of electrical impulses about the unique physical characteristics of a particular sound and this information is sent to the brain. After several relays within the brain, the messages finally reach the auditory areas of the cerebral cortex which process and interpret these signals as a musical phrase, a human voice or any of the range of sounds in the world around us at any particular moment.

There are about 16 000 hair cells in the human cochlea, compared to 100 million photoreceptors in the retina of the eye, and they are extremely vulnerable. Disease, ageing and over-exposure to loud noise, all take their toll on hair cells and by the age of 65, people tend to have lost 40% of them. Once destroyed, they do not regenerate.

This block, together with Chapters 1 to 6 of the Reader, examines the basic mechanisms responsible for our ability to hear. We begin by describing the nature of sound and then go on to consider how a sound stimulus is processed by the auditory system. We concentrate mainly on the cochlea, as it is here that the mechanical stimulus is transduced into an electrical response that can be forwarded to the brain for interpretation. We also examine the central auditory nervous system pathways and describe the physiological mechanisms responsible for our perception of pitch and loudness and our ability to localize the source of a sound stimulus.

We also examine, in less detail, a second function of the ear, namely how the ear enables us to answer two main questions basic to human life: ‘Which way is up?’ and ‘Where am I going?’ It does so by measuring linear and angular acceleration of the head through an ensemble of five sensory organs in the inner ear known as the vestibular system. This is described in Section 6 of the block, and Chapter 7 of the Reader.

Sound: the signal

2

2.1 What is sound?

The word 'sound' can have two meanings. It can be used to describe the signal, as when one talks about sound waves, or it can be used to describe the sensation. This section is going to look at the signal itself in preparation for the later sections that will look at the sensation of sound.

So what is sound? As you may recall from Block 1, sound originates from the motion or vibration of an object. A tuning fork produces a sound if its prongs are squeezed together and then released. (Striking the tuning fork against a solid object has the same effect.) The movement of the prongs causes movement of the air close to their surfaces. The air molecules alternately bunch together and move apart. When the molecules bunch together, local air pressure increases and when they are moved apart it decreases, as shown in Figure 2.1a (overleaf). Because the tuning fork is vibrating regularly, the result is a regular pattern of pressure variations called compressions and rarefactions, which move away from the tuning fork. If the increases and decreases in air pressure were measured at some fixed point in time, they would show a pattern similar to that shown in Figure 2.1b. The plot of pressure variation against distance suggests that it is a sinusoidal (or sine) wave. In this plot, the waveform repeats itself over a distance, marked as λ, known as the **wavelength**. The wave depicted in Figure 2.1b will also vary with time. Imagine a small floating object such as a fishing bob placed on the water. If the position of the bob is plotted against time then the plot might look similar to that shown in Figure 2.1c. In this case, the wave is characterized by the **period** (T), which is the time between successive peaks or troughs. Sounds that consist of a single, repeating sine wave like that shown in Figure 2.1 are known as **pure tones**.

Pure tones, represented by a sine wave, can be considered the basic element of more complex sounds and there are an infinite number of them. Because of this, we need to specify what makes one sine waveform different from another. There are in fact three basic characteristics by which our ears differentiate between different sinusoidal waveforms. One crucial way in which they differ is in the time it takes them to repeat. Figure 2.2 (overleaf) shows two waves that differ in this way. You can see that the lower wave takes longer to repeat than the upper one. The shape that repeats is called a **cycle** and the time it takes to repeat, as we have just seen, is the period. The period can be expressed in any units of time such as seconds or milliseconds. The same information can be specified by the **frequency** with which the cycle repeats. The frequency of a sound refers to the number of cycles that occur in one second. The unit of frequency is the **hertz** (**Hz**) which is just another name for cycles per second. Perhaps a more familiar term that you may have come across is the pitch of a sound. There is a relationship between pitch and frequency in that, in general, the higher the frequency of a sound, the higher its pitch and vice versa. However, as you will learn later, this relationship is not a simple linear one.

The term 'sound' is usually applied to any sound within the frequency range that is audible to humans. This is normally from about 20 Hz to 20 000 Hz in young people, but, as you will learn later, the top of the range decreases with increasing age. Sound with frequencies above 20 000 Hz are known as **ultrasonic**, those with frequencies below 20 Hz as **infrasonic**.

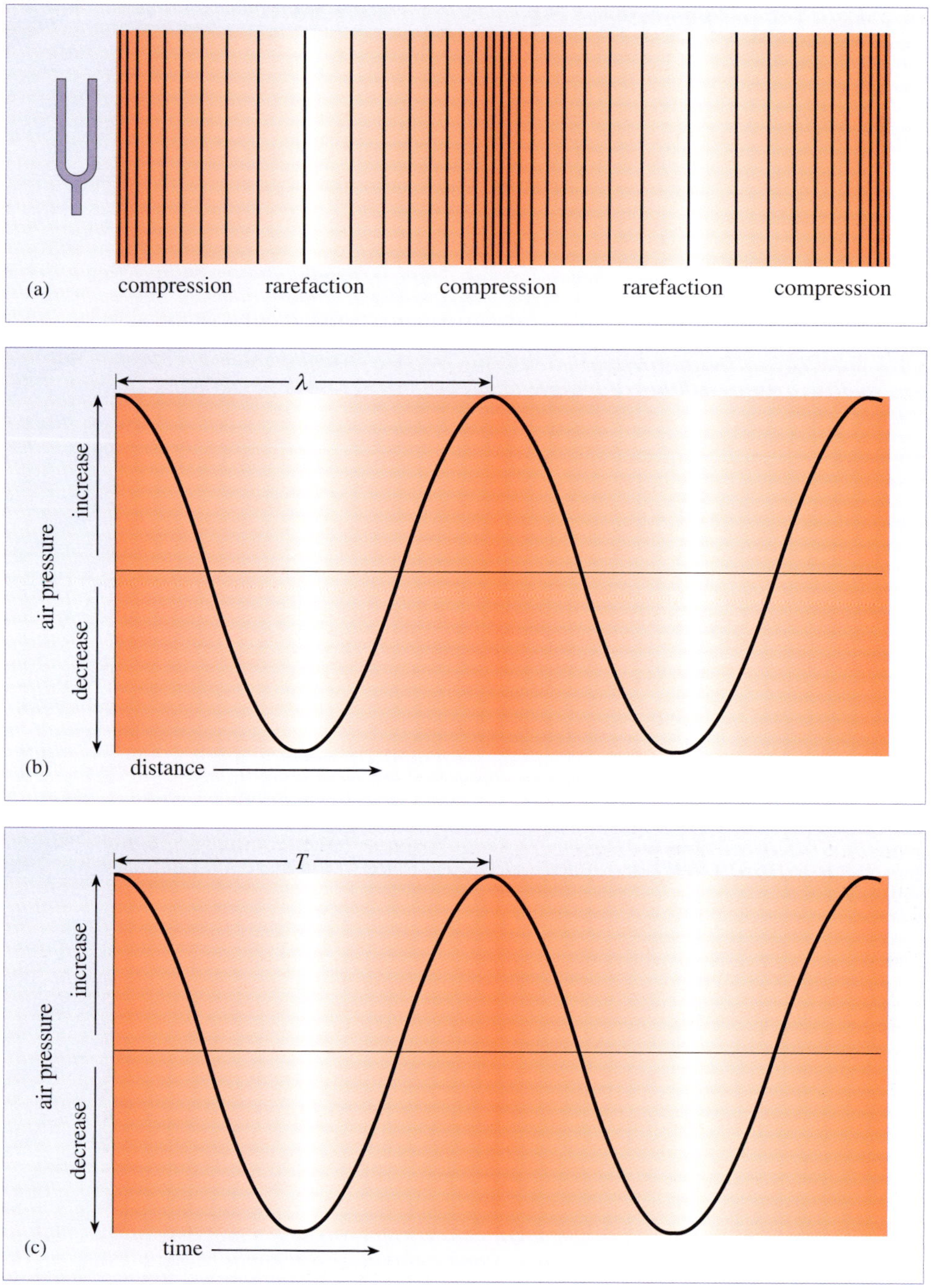

Figure 2.1 (a) The vibrations of the tuning fork produce a pressure wave moving away from the vibrating surface. (b) A plot of the change in pressure (from atmospheric pressure) versus distance at a given time. (c) A plot of the change in pressure versus time at a particular location.

A second characteristic of sinusoidal waves is that they differ in how big they are – their **amplitude**. The two waves in Figure 2.3 have the same frequency but look different. The lower wave has a greater amplitude than the upper one. The difference in amplitude between two waves can be quantified by measuring and comparing the height of the peaks. Differences in amplitude of auditory signals arise from differences in the displacement of the vibrating object. So, the amplitude of a sinusoidal wave produced by a tuning fork will depend on the force with which the prongs are 'twanged' – the more force used, the greater the height of the peaks and the greater the amplitude of the wave produced. Amplitude is a measure of local

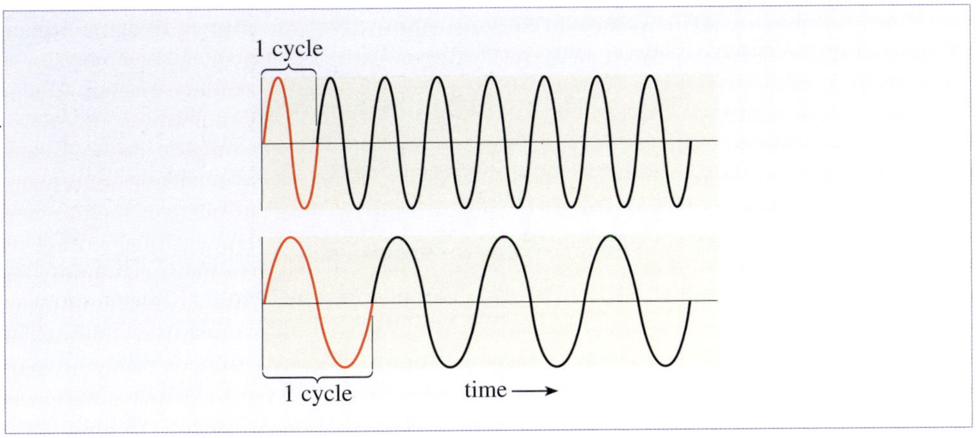

Figure 2.2 Two sinusoidal waves that differ in frequency. The time taken for one cycle to repeat (the period) is longer in the lower trace compared to the upper trace. This means that the lower trace has a lower frequency (i.e. has fewer cycles per second). These and the following waveform plots are equivalent to pressure (relative to atmospheric, in arbitrary units) versus time.

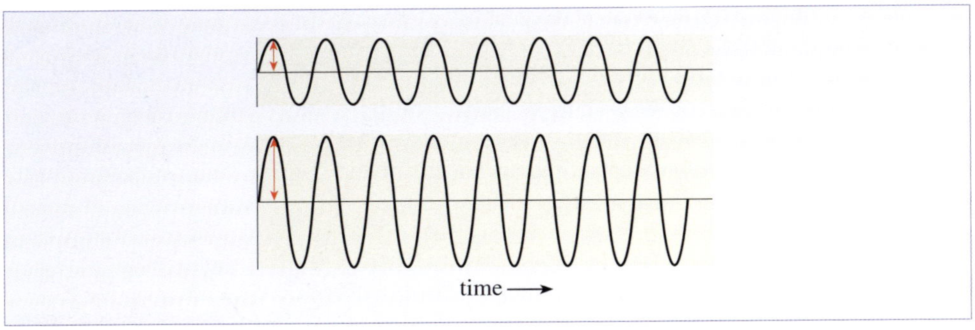

Figure 2.3 Two sinusoidal waves that differ in amplitude. The lower trace has a greater amplitude than the upper trace. (The amplitude is indicated by the red arrows.)

pressure relative to atmospheric pressure. Points above the line in Figure 2.3 represent an increase in pressure relative to ambient while points below the line represent pressure below atmospheric pressure. The amplitude of a sound is a major determinant of how loud or soft we perceive a sound to be.

There is a third way in which sinusoidal waves differ from one another. You can see in Figure 2.4 that both waves have the same frequency and amplitude, yet they are not identical. The difference is that the waveforms start at different points in their cycles. This is known as a difference in **phase**. In the top trace the waveform starts at zero and decreases, while in the bottom trace it starts at zero and increases. The quantity we use to specify differences in phase between waves is number of degrees. There are 360 degrees in one full cycle so the two waves in Figure 2.4 are 180 degrees out of phase. Waves of the same phase are said to be in phase.

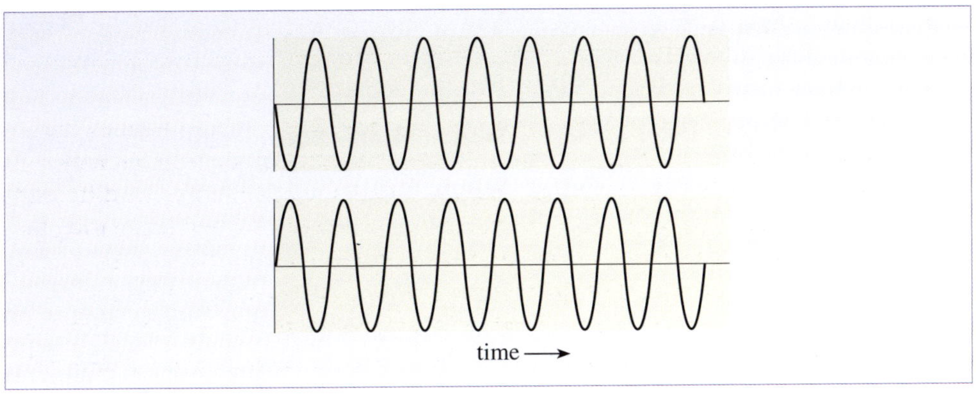

Figure 2.4 Two sinusoidal waves that differ in phase by 180 degrees.

- ❍ If two sinusoidal waves differ in phase by 360 degrees are they in phase or out of phase?
- ● They are in phase even though they are one full cycle apart.

Any sinusoidal wave (or pure tone) can therefore be uniquely specified by its frequency, its amplitude and its phase.

Sound waves in air and liquid are known as **longitudinal waves** because the movements of the molecules of air (or other medium) are in the direction of propagation of the wave. (The other possible kind of wave is a **transverse wave**. In a transverse wave the movement of the particles is perpendicular to the direction of propagation, as, for example, in the case of waves on the surface of water.)

Sound does not only travel through air but also through various other media. The speed of a sound wave, that is to say the speed with which sound energy can be transferred from one point to another, is different in different media and depends on the density of the medium and on its ability to be deformed. This should not be confused with the speed of motion of the particles in the medium, which is invariably much less than the speed of propagation of the wave. Table 2.1 gives the speed of sound in various materials. Note that the speed is greater in liquids than in gases and greater still in most solids.

Table 2.1 The speed of sound in different materials.*

Substance	Speed of sound / m s^{-1}	Temperature / K
air	343	293
water	1480	293
soft tissue	1540	–
bone	4080	–

*The temperature in air and water is included as the speed is temperature dependent. (In a gas it varies as $\sqrt{T}$ where T is temperature in Kelvins.)

2.2 Amplitude and intensity

The amplitude of a sound wave is expressed in terms of pressure. For sound, changes in air pressure are measured relative to atmospheric pressure and the SI unit for pressure is the pascal (Pa), where 1 Pa = 1 N (newton) m^{-2}. Often the pressures are so small that the micropascal (μPa = 10^{-6} Pa) is used.

However, when talking about sounds we rarely talk about the amplitude of a sound. Rather, we talk about sound **intensity**. For any wave, intensity is proportional to amplitude squared, so, for a sound wave we can write:

$$I \propto P^2 \qquad (2.1)$$

where I is the intensity and P is the pressure relative to atmospheric pressure. Therefore, a pressure measurement can be converted to an intensity measurement by squaring, and if we want to convert an intensity measurement to an amplitude measurement we can take the square root. The reason for doing these conversions is that although we speak about the intensity of a sound, most acoustic instruments measure pressure. The SI unit used to measure intensity is watts per unit area (normally $W\,m^{-2}$).

Our auditory system operates over an enormous range of intensity. If we were to assign one unit of intensity to the least intense sound that can be detected by the ear, then the loudest sound that we could tolerate before the ear was damaged would be 100 000 000 000 000 (10^{14}) units of intensity. The auditory system would therefore have a dynamic range of 10^{14} intensity units. This range is so large that it would be impossible to work with in practical situations (imagine trying to plot a graph from 1 to 10^{14} in 10 unit steps). Reducing the dynamic range to more manageable numbers is therefore necessary and this is where decibels come in. When intensity is considered on a ratio scale, logarithms can be used to describe the dynamic range of the auditory system (Box 2.1). (If you have not used logarithms before you should read through this section with the aim of understanding the ideas. Don't worry about the formulae and calculations.)

The starting point is the logarithm of the ratio of two intensities, I_1 and I_2. This logarithm is labelled a bel but because a bel is too large an interval to work with, a **decibel** (abbreviated dB: one-tenth of a bel) is used. The ratio of two intensities expressed in decibels is:

$$\text{dB} = 10 \log_{10}(I_1 / I_2) \quad (2.2)$$

where intensity I_1 is being compared with intensity I_2.

From now on, since we shall only be concerned with logs to the base 10, we shall drop the subscript 10.

So if I_1 is equal to 10^{14} and I_2 is equal to 1, then:

$$10 \log(10^{14}/1) = 10 \log 10^{14} = 10 \times 14 = 140 \text{ dB}$$

The dynamic range of the auditory system can now be expressed in terms of decibels, yielding a much more convenient range of numbers to work with. The original interval scale of intensities ranging from 1 to 10^{14} has been reduced to a new interval scale of decibels ranging from 0 to 140 by transforming intensity units to decibels using logs.

The formula for decibels gives the relative *intensity* of a sound in dB (in this case the most intense sound we can tolerate is 140 dB above the sound that we can only just hear). Sometimes, as already mentioned, it is necessary to measure the *amplitude* (that is the change in pressure P), rather than the intensity of a sound and because
$I \propto P^2$, it follows that:

$$\text{relative intensity} = 10 \log(I_1/I_2) = 10 \log(P_1/P_2)^2 = 20 \log(P_1/P_2) \quad (2.3)$$

It is important to remember that the decibel is the ratio of two quantities. This means that saying something has an intensity or amplitude of 60 dB is meaningless. This statement does not indicate whether the intensity is 60 dB above the most intense sound you can tolerate, 60 dB below the lowest sound you can hear, 60 dB above the sound that your friend can hear, or anything else. When talking about the level of a sound we therefore need a **reference pressure** (P_0) or **reference intensity** (I_0). The hypothetical reference intensity we used in the example above was 1 unit of intensity. The reference intensity commonly used in acoustics is $10^{-12}\,W\,m^{-2}$ which is the lowest intensity of a 1000 Hz tone that an average young adult can detect.

Box 2.1 Logarithms

Logarithms (logs) are a very useful concept. If you have not used them for a while this box will provide a reminder.

The log of a number (e.g. the log of a) is defined as the power to which you have to raise the base b to get the number a. That is:

if $a = b^x$

then $\log_b a = x$

Note that the base is written as a subscript to the word 'log'. Although it is possible to write logs to *any* base, logs to the base 10 are perhaps the most commonly used. For these we would say that:

if $a = 10^x$

then $\log_{10} a = x$

Here are a couple of examples:

since $1000 = 10^3$, it follows that $\log_{10} 1000 = 3$

since $0.1 = 10^{-1}$, it follows that $\log_{10} 0.1 = -1$

Note that logarithms and exponents are *inverse* operations. If you use your calculator to determine $\log_{10} a$, then inputting 10 to the power of the answer will return the original value a. Try it!

Using logs

There are some properties of logs that make them very easy to handle. The two most useful ones are probably:

$$\log(q \times r) = \log q + \log r$$
$$\log(q / r) = \log q - \log r$$

i.e. if the numbers are multiplied, the logs are added, or if the numbers are divided, the logs are subtracted.

From these rules, two other rules follow, namely that:

$$\log a^n = n \log a$$

(since $\log a^n = \log(a \times a \times a \times \ldots) = \log a + \log a + \log a + \ldots = n \log a$)

and also that:

$$\log(a / b) = -\log(b / a)$$

(Since $\log(a / b) = \log a - \log b = -(\log b - \log a) = -\log(b / a)$)

10^{-12} W m^{-2} corresponds to a pressure of 20 μPa. When decibels are expressed relative to a reference pressure of 20 μPa, they are expressed in terms of **sound pressure level** (**SPL**). Hence a sound of 60 dB SPL, is 60 dB above the reference level of 20 μPa. When decibels are expressed relative to 10^{-12} W m^{-2} they are expressed in terms of **intensity level**. Thus 'level' refers to a dB measure.

Table 2.2 gives some typical values.

Table 2.2 Intensity levels in decibels SPL for some familiar sounds.

Sound	Intensity in dB SPL
jet at take off (generally above the pain threshold)	140
loud rock music	120
shot gun	100
door slamming	80
normal conversation	60
muffled conversation	40
soft whisper at 1.5 m	30
virtual silence	10
threshold of hearing	0

Continuous noise of 90 dB SPL or over is potentially hazardous. You will discover why when you read Chapter 6 of the Reader.

❍ If the pressure level of a sound is 20 μPa (which is the standard reference pressure used in acoustics), what is its intensity level in dB SPL?

● $\text{dB} = 20 \log (20/20)$

$$= 20 \log 1$$
$$= 20 \times 0$$
$$= 0 \text{ dB SPL}$$

In other words, 20 μPa (or 10^{-12} W m^{-2}) corresponds to an SPL of 0 dB.

You should note that doubling the amplitude of a sound does not double its sound pressure level but increases it by 6 dB. So, if we have a sound with an SPL of 20 dB and we double its amplitude, the sound pressure level does not double to 40 dB. Rather, the SPL increases from 20 to 26 dB as shown below.

If the pressure level of a sound is 200 μPa, then:

$$\text{dB} = 20 \log (200/20)$$
$$= 20 \log 10$$
$$= 20 \text{ dB SPL}$$

If the pressure level of the sound is doubled to 400 μPa, then:

$$\text{dB} = 20 \log (400/20)$$
$$= 20 \log 20$$
$$= 20 \times 1.3$$
$$= 26 \text{ dB SPL}$$

We have already mentioned that a wave's amplitude is a major determinant of perceived loudness. However, as you will see later, other factors such as frequency also affect our experience of loudness. For example, as we have shown above, a 26 dB SPL sound is double the amplitude of a 20 dB SPL sound but it is not double its loudness. The perception of loudness will be dealt with later in this block.

2.3 Analysis of sounds

Until now we have been discussing sound waves that take the form of a pure sinusoidal wave. In reality however, most sound sources are made up of a mixture of different frequencies and the waveform is more complex. For example, Figure 2.5 shows the sound waves produced by various different sources. In the case of the musical instruments and the human voice, both of which are making a continuous

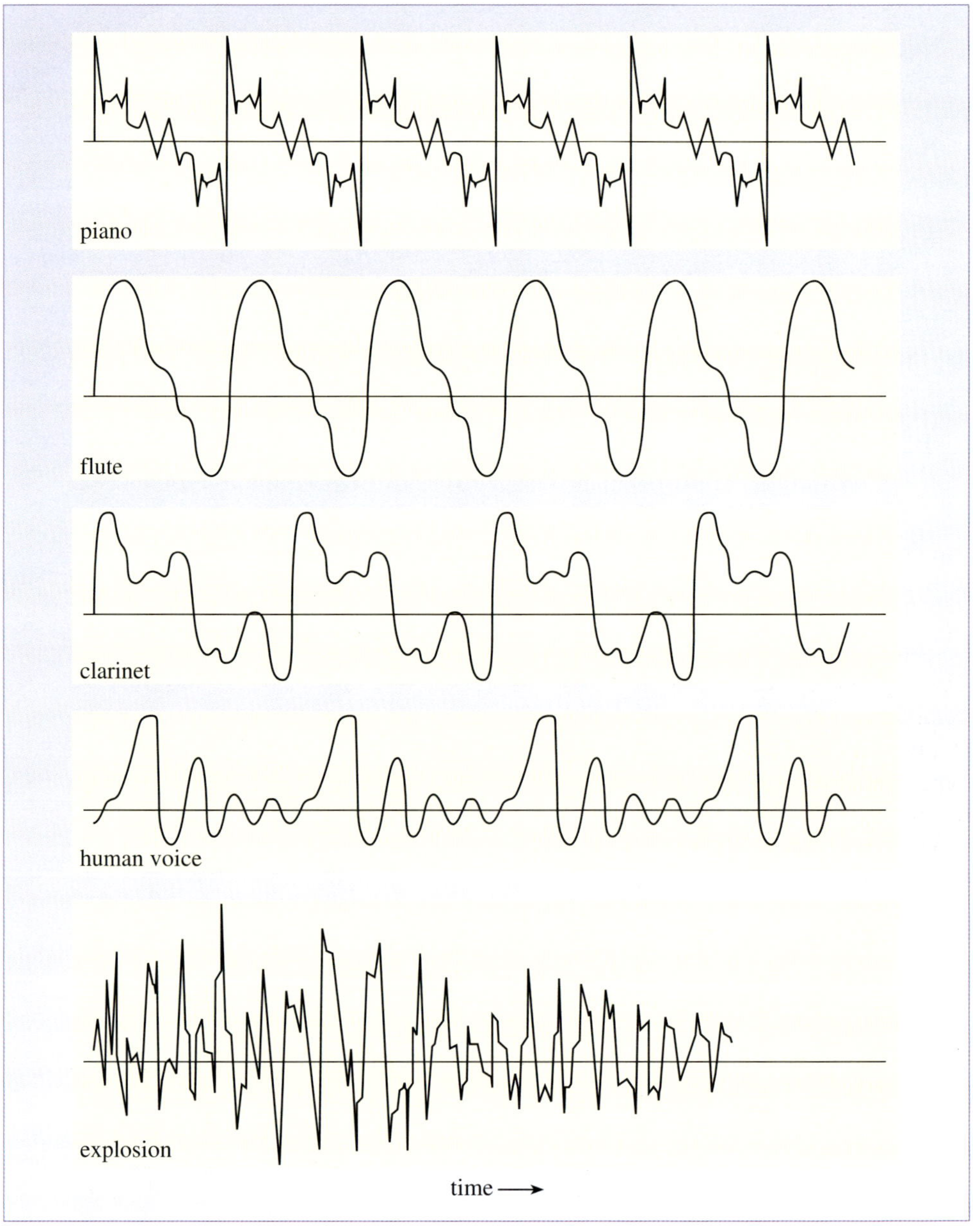

Figure 2.5 The sound waves produced by various different sources.

sound (a single note in the case of the musical instruments and a sound like a hum or a vowel sound in the case of the human voice), it is clear that the pattern repeats itself – this is a **periodic waveform**.

A complex waveform like this can be obtained by adding several simple, sinusoidal, waves. This is demonstrated in Figure 2.6 using two sound waves, plotted as pressure variation against time.

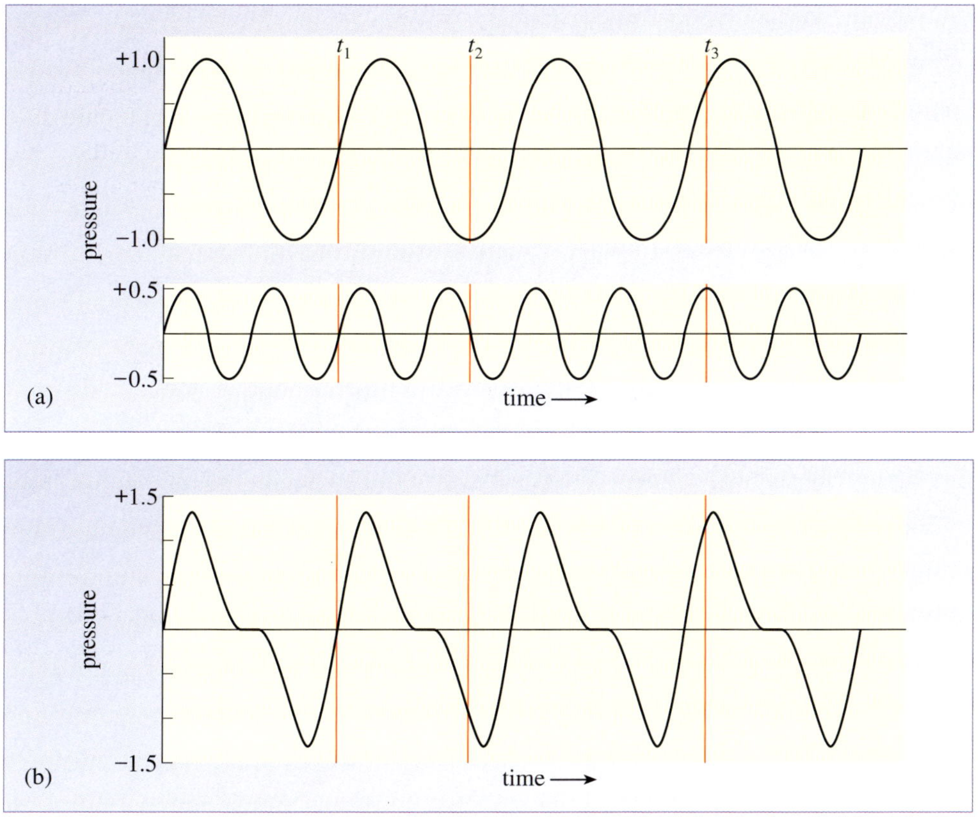

Figure 2.6 The addition of two waveforms, one with an amplitude twice that of the other but half the frequency (a), gives the resultant waveform (b).

Basically, adding any two waveforms simply involves adding their respective values at each point in time and plotting the result as a function of time. In Figure 2.6a, the amplitude of the higher frequency sine wave (lower figure) is half that of the upper. At any point in time, we know the pressure due to each of the sine waves from the graph. So, at time t_1 in the figure, both waveforms have a value of 0. Therefore if the two waves are added together at time t_1, the value of their sum is 0. At time t_2, the value of the upper wave is –1 and that of the lower is 0. So, if they are added together at this point in time, the overall pressure would be:

$$-1 + 0 = -1$$

❍ What would the value of the resultant wave be at time t_3?

● At time t_3, the value of the upper wave is 0.7 and the value of the lower wave is 0.5. So, by adding them together we get $0.7 + 0.5 = 1.2$

Just as we can add sinusoidal waves to generate a complex wave, we can also decompose a complex wave into its component sinusoidal waves. This process of breaking a periodic waveform down into its component frequencies is known as

Fourier analysis after the French mathematician Joseph Fourier (1768–1830). Fourier's theorem states that a periodic waveform can only contain sinusoidal waves that are **harmonically related** to one another. Harmonically related means that the frequency of a sinusoidal wave is a multiple of some 'basic' or 'fundamental' number. For example, if a waveform has a period of 10 ms (repeats itself every 10 ms) its fundamental frequency is 100 Hz (since it repeats 100 times every second: 1 s = 1000 ms) and its harmonically-related frequencies are 200 Hz, 300 Hz, 400 Hz, etc. Each of these component frequencies is known as a **harmonic**. This sequence of related harmonics is known as a **harmonic series**. Thus the analysis of a periodic waveform into its component sinusoidal waves is referred to as calculating its **Fourier series**. Figure 2.7 shows a sawtooth wave that is periodic as it repeats itself over and over again with a period of 10 ms.

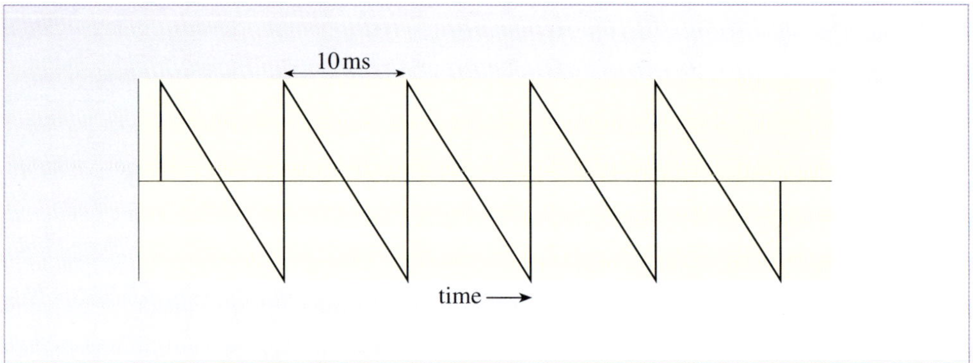

Figure 2.7 A sawtooth wave with a period of 10 ms and a frequency of 100 Hz.

This wave was generated by simply adding together a number of sinusoidal waves that are multiples of 100 Hz. Figure 2.8 demonstrates how this is achieved. Note that the amplitudes of the component waves differ (in fact, amplitude decreases as the harmonic number increases).

Fourier analysis decomposes the wave into its constituent frequencies which can then be described by its **frequency spectrum**. Amplitude is drawn on the vertical axis and frequency on the horizontal axis. A vertical line is then drawn at the frequencies where the harmonics are present in the signal. The height of the line is proportional to the amplitude. The amplitude of the first 10 components of a sawtooth wave would look like that shown in Figure 2.9 (overleaf). This process can also be carried out on a waveform that is aperiodic (the waveform does not repeat), such as speech, in which case the equivalent is a continuous curve and is referred to as a **Fourier transform**.

The ear, in fact, functions much like a Fourier analyser by decomposing complex sounds into their component frequencies. Fourier analysis is also important in vision but in a rather different way as you will see in Block 4.

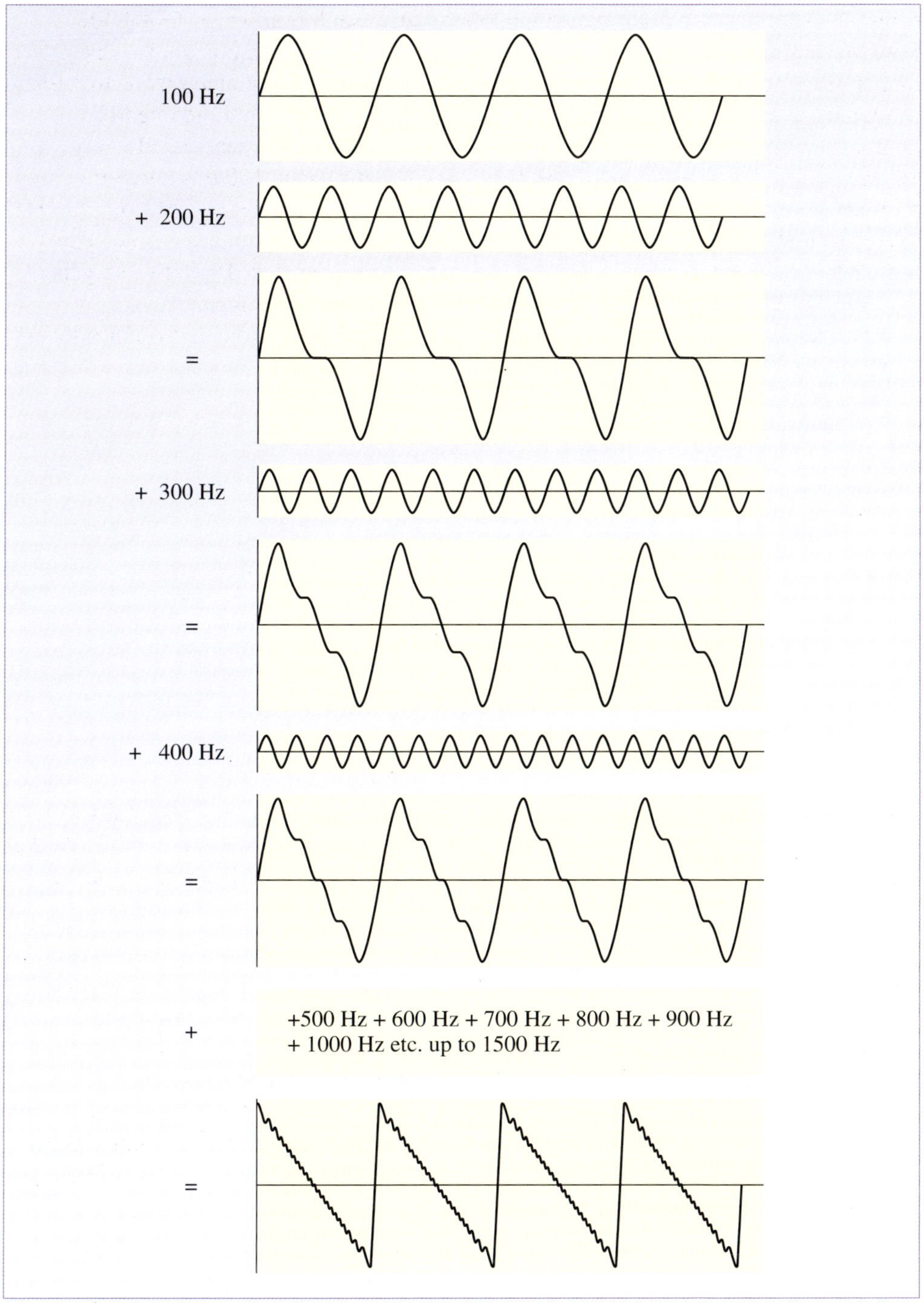

Figure 2.8 The synthesis of a sawtooth wave. The bottom trace shows the waveform after the addition of 15 terms of the Fourier series. The approximation to the sawtooth wave seen in Figure 2.7 is fairly close, although an infinite number of terms are needed to produce an exact match.

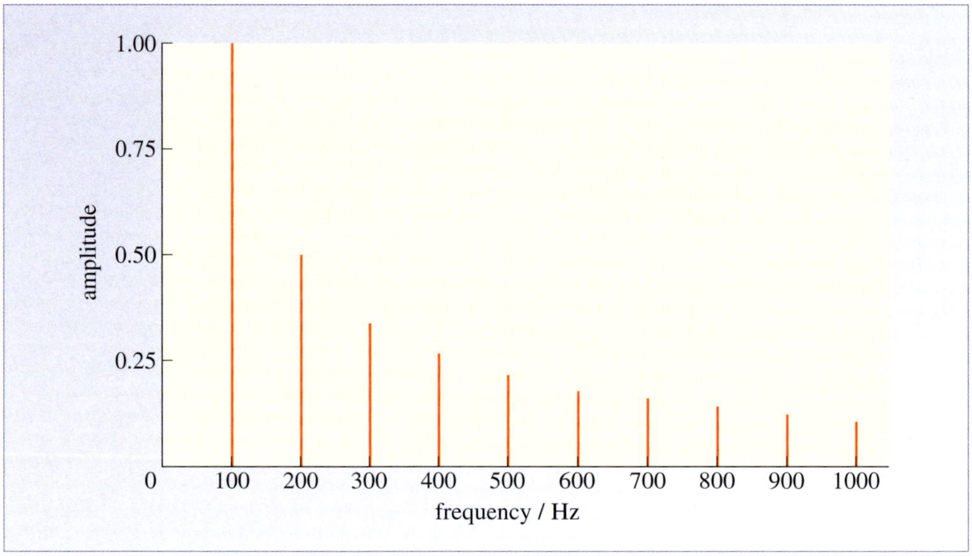

Figure 2.9 The first 10 component waves of the Fourier series of a sawtooth wave. The amplitude of the components drops off with increasing frequency.

2.4 Filters

Now that the idea of a frequency spectrum (or Fourier transform) has been introduced it is appropriate to discuss **filters**. The use of filters is very widespread – they turn up in sound and vision as well as in other fields such as electronics and communications.

The idea of a filter is to remove some of the frequencies in a signal and allow others to pass through. This of course affects the frequency spectrum. There are four important classes of filter: low-pass, high-pass, band-pass and notch.

A **low-pass filter**, as the name suggests, allows the low frequencies through but rejects the high frequencies. The effect of a low-pass filter is shown in Figure 2.10a. Ideally all frequencies above the cut-off frequency will be lost and all frequencies below will be retained (dotted line); however, in practice the filter will not be perfect (solid line). One reason for using a low-pass filter might be to cut out high-frequency noise.

A **high-pass filter** does the opposite and the action of a perfect high-pass filter on the same waveform is shown in Figure 2.10b. High frequencies are retained and low ones lost.

A **band-pass filter** allows a limited range (or band) of frequencies through. This range depends on the lower and upper limits of the band. Band-pass filters are rarely 'perfect' (dotted line) and are likely to have the more realistic profile shown by the solid line in Figure 2.10c. Band-pass filters are very relevant to hearing.

A **notch filter** is the opposite of a band-pass filter. It cuts off frequencies in a certain range and allows the frequencies above and below the band to pass through.

STUDY FILE

Activity 2.1 Sound waves

You should now undertake this CD-ROM activity, which is in two parts. The first examines the properties of sound waves produced by a tuning fork; the second is a demonstration of Fourier analysis. Further instructions are given in the Block 3 *Study File*.

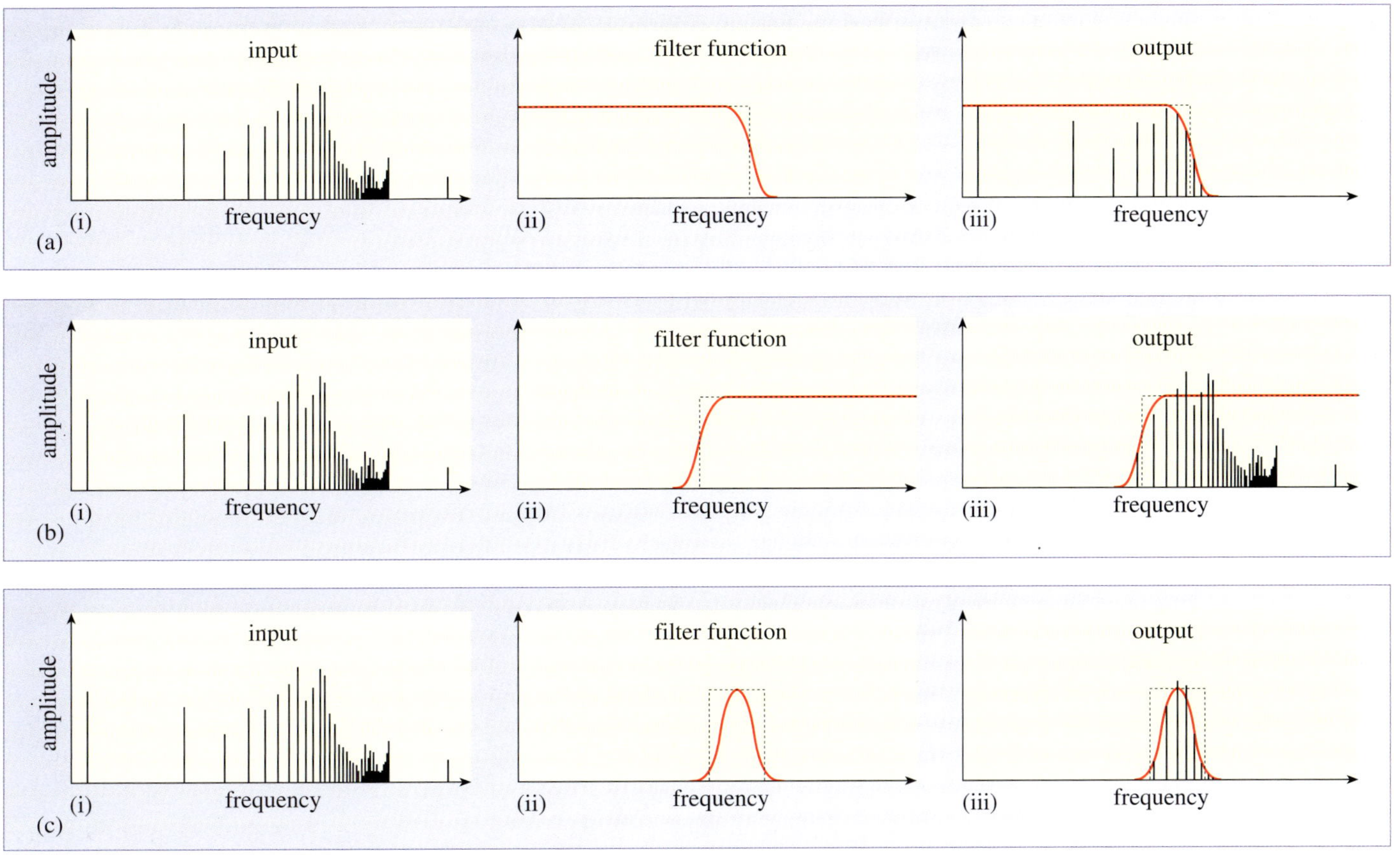

Figure 2.10 Three classes of filter. For each filter, the new frequency spectrum is the product of the original spectrum and the filter function. The dotted line on the filter function and on the final spectrum represents the ideal filter, the solid line the realistic one. (a) A low-pass filter; (b) a high-pass filter; (c) a band-pass filter. The fall-off on either side of the central frequency is likely to be gradual (solid line).

Summary of Section 2

Sound is produced by the motion or vibration of an object. Sound waves are longitudinal pressure waves, which travel at different speeds in different media.

The simplest sounds are composed of single sinusoidal waves, which are known as pure tones and can be characterized by their frequency, amplitude and phase. Frequency is measured in hertz, amplitude in dB SPL and phase in degrees. The range of frequencies audible to the human ear is approximately 20–20 000 Hz.

The intensity of a sound is proportional to the square of its amplitude (pressure change). The decibel scale is used to describe the relative amplitude or intensity of a sound. Sound levels are usually compared with the threshold of hearing of the normal ear.

Complex sounds are composed of a number of simple sinusoidal waves added together. Periodic waveforms repeat indefinitely while aperiodic waveforms do not.

An analysis of a periodic signal into its component waves is referred to as calculating its Fourier series. When aperiodic signals are characterized by their frequency content, the process is referred to as calculating a Fourier transform.

Sounds can be filtered. A low-pass filter allows low frequencies to pass but removes high frequencies. A high-pass filter does the opposite. A band-pass filter allows through frequencies that are within a certain range and a notch filter does the opposite.

Question 2.1

What is the approximate wavelength of the note corresponding to middle C (262 Hz) (a) in air and (b) in water?

Question 2.2

What is the advantage of using the decibel scale?

Question 2.3

How many decibels correspond to a reduction in intensity by a factor of two?

Question 2.4

Draw sketches (similar to Figure 2.10b and c) to show the effect of a notch filter on the frequency spectrum in Figure 2.10a.

Sound reception: the ear

3

In order to hear a sound, the auditory system must accomplish three basic tasks. First it must deliver the acoustic stimulus to the receptors; second, it must transduce the stimulus from pressure changes into electrical signals; and third, it must process these electrical signals so that they can efficiently indicate the qualities of the sound source such as pitch, loudness and location. How the auditory system accomplishes these tasks is the subject of much of the rest of this block. We will begin by describing the basic structure of the ear, which carries out the first of the three tasks.

The human ear can be divided into three fairly distinct components according to both anatomical position and function: the **outer ear**, which is responsible for gathering sound energy and funnelling it to the eardrum; the **middle ear** which acts as a mechanical transformer; and the **inner ear** where the auditory receptors (hair cells) are located.

3.1 Structure and function of the outer and middle ears

Figure 3.1 is a diagram of the human ear. The outer ear consists of the visible part of the ear or **pinna**, the **external auditory canal (meatus),** and the **tympanic membrane (tympanum)** or eardrum. The human pinna is formed primarily of cartilage and is attached to the head by muscles and ligaments. The deep central portion of the pinna is called the **concha**, which leads into the external auditory canal, which in turn leads to the tympanic membrane.

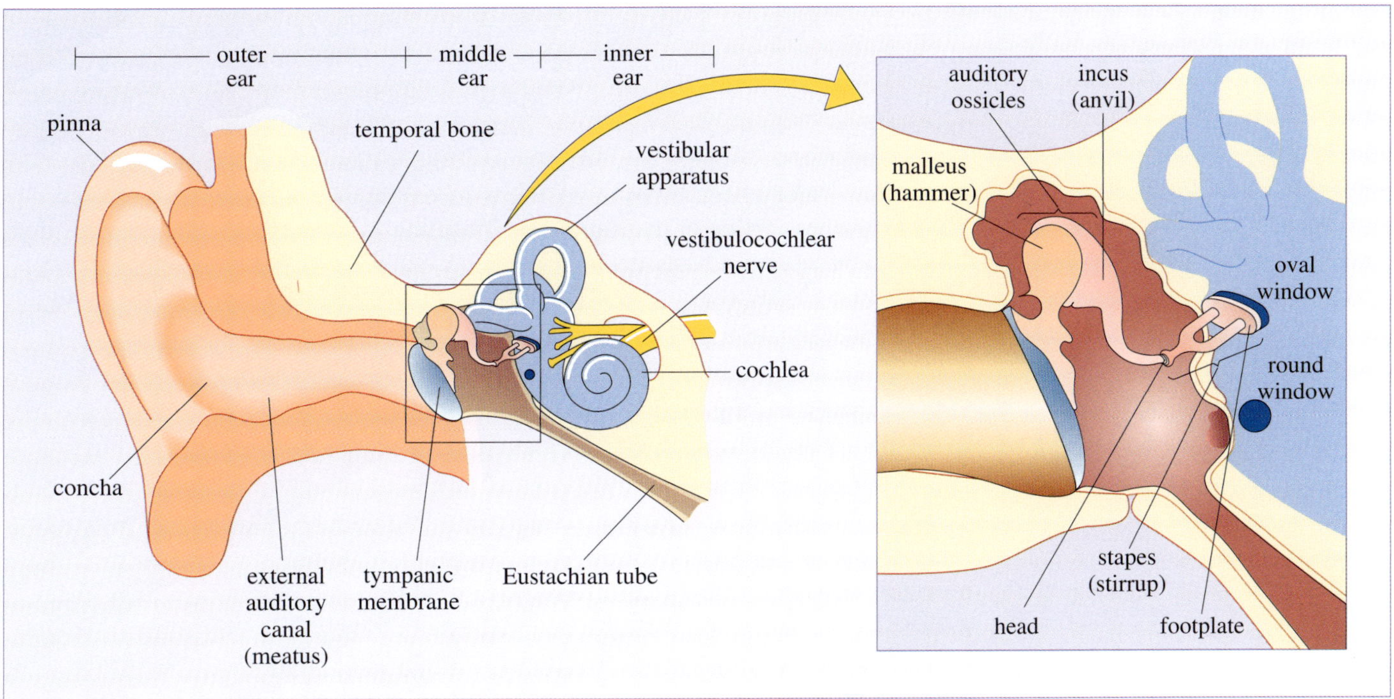

Figure 3.1 The human ear showing the outer, middle and inner ears.

Only mammals have pinnae and only some have mobile pinnae. The pinnae of humans and primates have no useful muscles and are therefore relatively immobile. Mobile, and to some extent, immobile pinnae help in localizing sounds by funnelling them towards the external canal.

❍ How would the immobility of our pinnae affect how we localize a sound source?

● Unlike animals with mobile pinnae, we must reposition our head in order to aim our ears at a sound source.

The pinnae also help in distinguishing between noises originating in front of and behind the head, and in providing other types of filtering of the incoming sound wave. In addition, the concha and external auditory canal effectively enhance the intensity of sound that reaches the tympanic membrane by about 10 to 15 dB. This enhancement is most pronounced for sounds in the frequency range of roughly 2 to 7 kHz and so, in part, determines the frequencies to which the ear is most sensitive. Finally, the outer ear protects the tympanic membrane against foreign bodies and changes in humidity and temperature.

The external auditory canal extends about 2.5 cm inside the skull before it ends in the tympanic membrane. Sound travels down the meatus and causes the tympanic membrane to vibrate. The tympanic membrane is thin and pliable so that a sound, consisting of compressions and rarefactions of air particles, pulls and pushes at the membrane moving it inwards and outwards at the same frequency as the incoming sound wave. It is this vibration that ultimately leads to the perception of sound. The greater the amplitude of the sound waves, the greater the deflection of the membrane. The higher the frequency of the sound, the faster the membrane vibrates.

On the other side of the tympanic membrane is the middle ear (Figure 3.1) which is an air-filled chamber containing three interlocking bones called **ossicles**. These are the smallest bones in the body and function to transmit the vibrations caused by auditory stimulation at the tympanic membrane to the inner ear. The bones are called the **malleus** (Latin for 'hammer'), the **incus** ('anvil') and the **stapes** ('stirrup'). The ossicle attached to the tympanic membrane is the malleus, which forms a rigid connection with the incus. The incus forms a flexible connection with the stapes. The flat bottom portion of the stapes, the **footplate**, is connected to the **oval window** (a second membrane covering a hole in the bone of the skull). In response to sound, the inward–outward movement of the tympanum displaces the malleus and incus and the action of these two bones alternately drives the stapes deeper into the oval window and retracts it, resulting in a cyclical movement of fluid within the inner ear.

This may seem a complex way to transmit vibrations of the tympanic membrane to the oval window. Why must they be transmitted via the ossicular chain and not simply transferred directly?

The reason is that the middle ear cavity is *air*-filled while the inner ear is *fluid*-filled. The passage of sound information from the outer to the inner ear involves a boundary between air and fluid. If you have tried talking to someone who is under water, you may have observed that sound does not travel efficiently across this kind of interface. In fact, 99.9% of the sound energy incident on an air/fluid boundary is reflected back within the air medium and only 0.1% is transmitted to the fluid. Therefore, if sound waves were to impinge directly on the oval window, the membrane would barely move. Most of the sound would be reflected back because the fluid in the inner ear is denser than air and resists being moved much more than air does. Consequently, in order to drive the movement of the oval window and vibrate the fluid, greater pressure is needed.

The middle ear provides two ways of doing this. The first is to do with the relative sizes of the tympanic membrane and the stapes footplate (which is connected to the oval window). Measurements have shown that the area of the tympanic membrane that vibrates in response to high intensity sound is 55 mm^2. The stapes footplate which makes contact with the oval window has an area of only about 3.2 mm^2. So, if all the force exerted on the tympanic membrane is transferred to the stapes footplate, then the pressure (force per unit area) must be greater at the footplate because it is smaller than the tympanic membrane. One rather painful demonstration of this principle is to compare the pressure exerted on your toe by someone wearing a stiletto heel compared to the pressure exerted by the same person wearing an ordinary trainer.

The second way in which the middle ear ossicles transfer the force from the tympanic membrane to the stapes footplate is through the lever action of the ossicles. Figure 3.2 shows how a lever system can increase the force of an incoming signal.

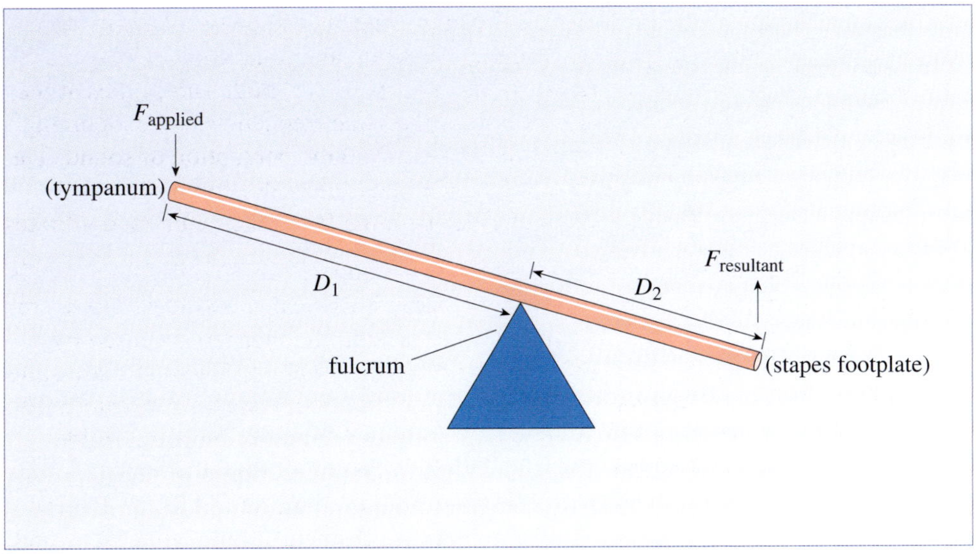

Figure 3.2 The lever action of the middle ear. The length of the malleus corresponds to D_1, the distance between the applied force and the fulcrum. The length of the incus corresponds to D_2, the distance between the fulcrum and the resultant force. If D_2 is less than D_1, then the resultant force will be greater than the applied force.

The middle ear has another function in addition to the mechanical transformation of the auditory signal. When the auditory system is subjected to very loud sounds that are potentially harmful to the inner ear, two set of muscles, the **tensor tympani** and the **stapedius muscles**, contract and in so doing reduce the magnitude of the vibration transmitted through the middle ear. The response of these muscles to loud noises is known as the **acoustic** or **middle ear reflex**.

The lever action of the middle ear and the middle ear reflex are covered in more detail in Chapter 1 of the Reader, which you will be asked to read at the end of Section 3.2.2.

3.2 The structure and function of the inner ear

The inner ear (Figure 3.3) can be divided into three parts: the **semicircular canals**, the **vestibule** and the **cochlea**, all of which are located in the temporal bone. The semicircular canals and the vestibule affect the sense of balance and are not concerned with hearing. However, the cochlea, and what goes on inside it, provides the key to understanding many aspects of auditory perception and will therefore be dealt with in some detail.

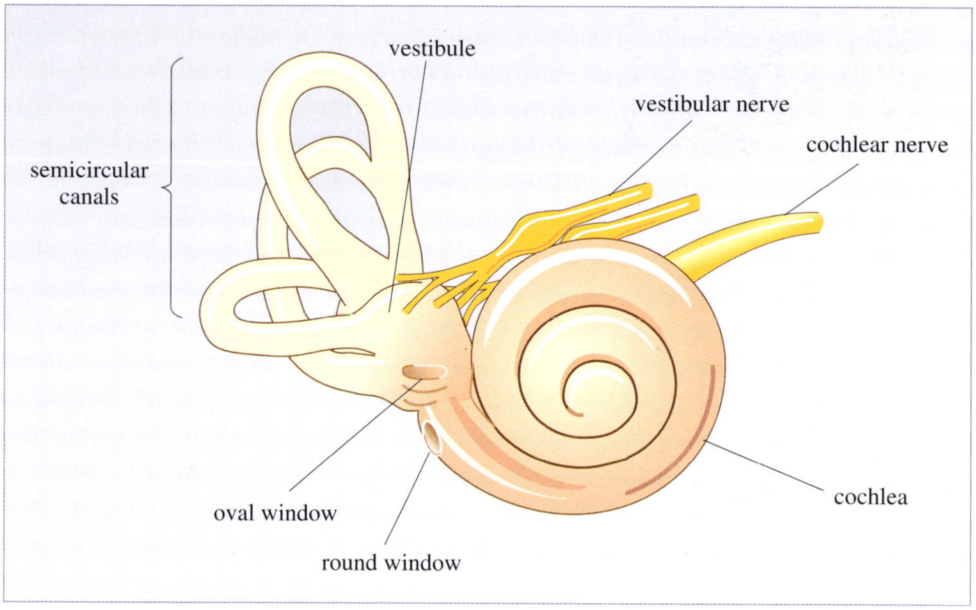

Figure 3.3 The inner ear showing the semicircular canals and the cochlea.

3.2.1 The anatomy of the cochlea

The cochlea has a spiral shape resembling the shell of a snail (Figure 3.4a). You can approximate the structure of the cochlea by wrapping a drinking straw 2.5 times around the tip of a sharpened pencil. The hollow tube, represented by the straw, has walls made of bone and the central pillar of the cochlea, represented by the pencil, is a conical bony structure called the **modiolus**. Unravelled (Figure 3.4b), the cochlea's hollow tube is about 32 mm long and 2 mm in diameter. The tube of the cochlea is divided into three chambers: the **scala vestibuli**, the **scala media** (or **cochlear duct**) and the **scala tympani**. The three scalae wrap around inside the cochlea like a spiral staircase ('scala' is Latin for 'stairway'). The scala vestibuli forms the upper chamber and at the base of this chamber is the oval window. The lowermost of the three chambers is the scala tympani. It too has a basal aperture, the **round window**, which is closed by an elastic membrane. The scala media or cochlear duct separates the other two chambers along most of their length. The start of the cochlea, where the oval and round windows are located is known as the **basal** end, while the other end, the inner tip is known as the **apical** end (or apex). The scala vestibuli and the scala tympani communicate with one another via the **helicotrema**, an opening in the cochlear duct at the apex. Both scala vestibuli and scala tympani are filled with the same fluid, known as **perilymph** (essentially the same in composition as the extracellular fluid bathing most of the nervous system), while the scala media is filled with **endolymph** (with very high potassium and low sodium concentrations).

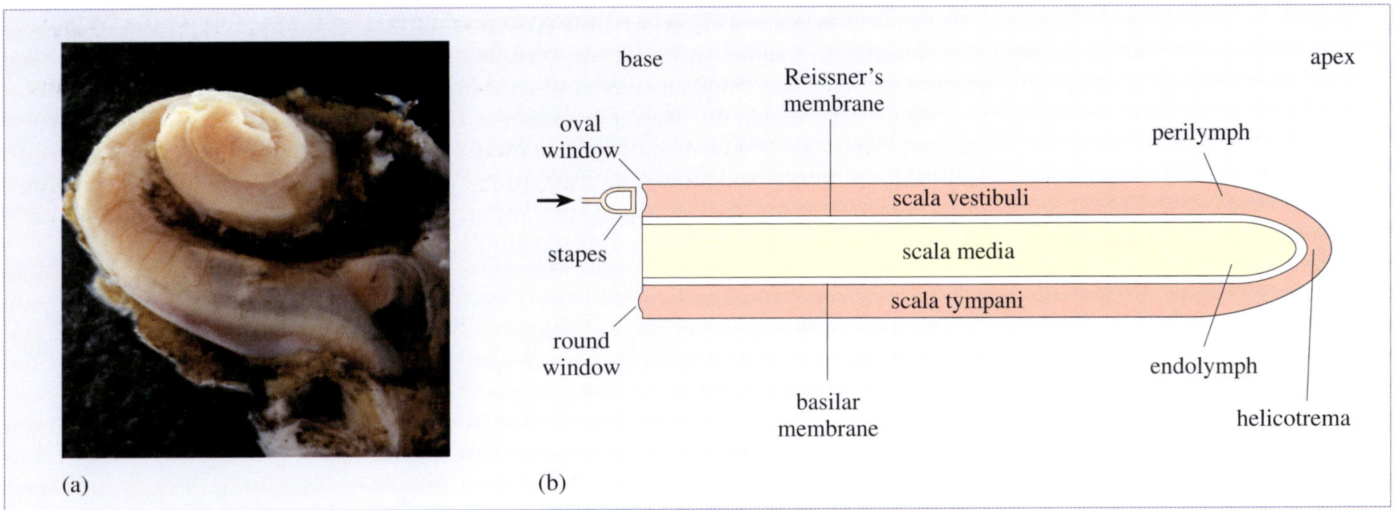

Figure 3.4 (a) Coiled cochlea of a human fetus of five months gestation. (Magnification × 40.) (b) Diagrammatic representation of the three scalae of the cochlea (uncoiled).

Figure 3.5 is a cross-section of the cochlea showing the three chambers which run along its length. Between the scala vestibuli and the scala media is a membrane called **Reissner's membrane** and between the scala tympani and the scala media is the **basilar membrane**. Lying on top of the basilar membrane within the cochlear duct is the **organ of Corti** and hanging over the organ of Corti, is the **tectorial membrane**. The collective term for the partitions of the scala media (the organ of Corti, the basilar membrane and the tectorial membrane) is the **cochlear partition**.

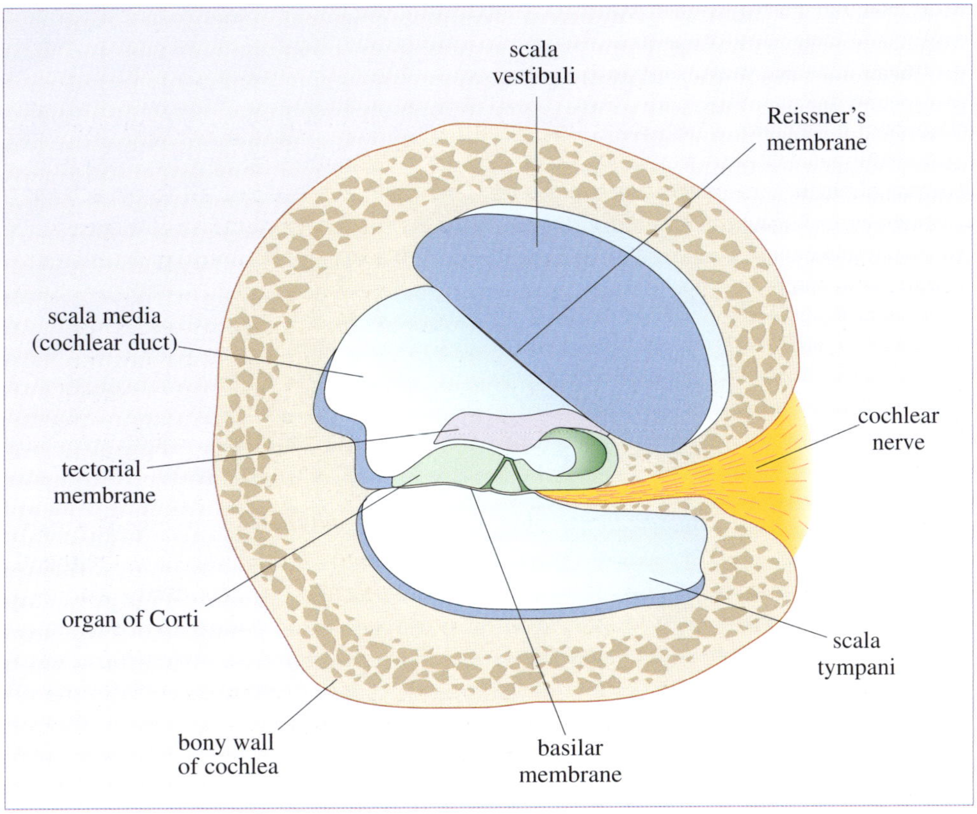

Figure 3.5 Cross-section of the cochlea.

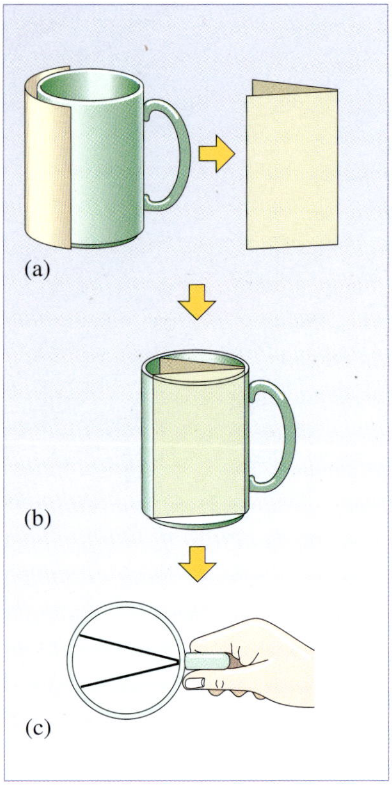

Figure 3.6 Construction of a 'cochlea'.

The following simple activity will help you understand the structure of the cochlea more clearly. Imagine that your empty coffee mug (assuming it's a cylindrical shape, or thereabouts) represents a section of the uncoiled tube-like cochlea. Take a piece of paper that is about the same height as the mug and that is wide enough to wrap halfway around the mug (Figure 3.6a). Fold it in half and insert it into the middle of the mug so that the fold runs vertically down the side of the mug where the handle is located (Figure 3.6b). The paper should lie across the middle of the mug. Now (if necessary) separate the two pieces of paper so that they form a V-shape (Figure 3.6c). You now have the basic structure of the cochlea. If you hold the mug by the handle in your right hand and look directly into it, you have the same view as that shown in Figure 3.5. Compare Figure 3.5 to your 'cochlea' and answer the following questions:

❍ What structure does the top piece of paper represent?

● Reissner's membrane.

❍ What does the space above the top piece of paper represent?

● The scala vestibuli.

❍ What does the bottom piece of paper represent?

● The basilar membrane.

❍ What does the space below the bottom piece of paper represent?

● The scala tympani.

❍ What does the V-shaped space represent?

● The scala media (or cochlear duct).

The scala media houses the organ of Corti. The organ of Corti and all its associated structures (including the hair cells, see below) runs the length of the basilar membrane (from the top to the bottom of your mug) as does the overlying tectorial membrane.

An enlargement of the organ of Corti is shown in Figure 3.7. The organ of Corti is the primary auditory receptor structure and houses the sensory receptor cells which are known as **hair cells** because each has about 100 hair-like **stereocilia** extending from its apical end which are embedded in the tectorial membrane. You can see that the hair cells are of two types: **outer hair cells** and **inner hair cells**, which are separated by a rigid inverted V-shaped structure known as **Corti's arch**.

We shall return to the organ of Corti and the hair cells and their involvement in the transduction of an auditory signal into neural information in Section 3.2.3. For now, we need to consider the basilar membrane and in particular, its response to vibrations in the cochlear fluid.

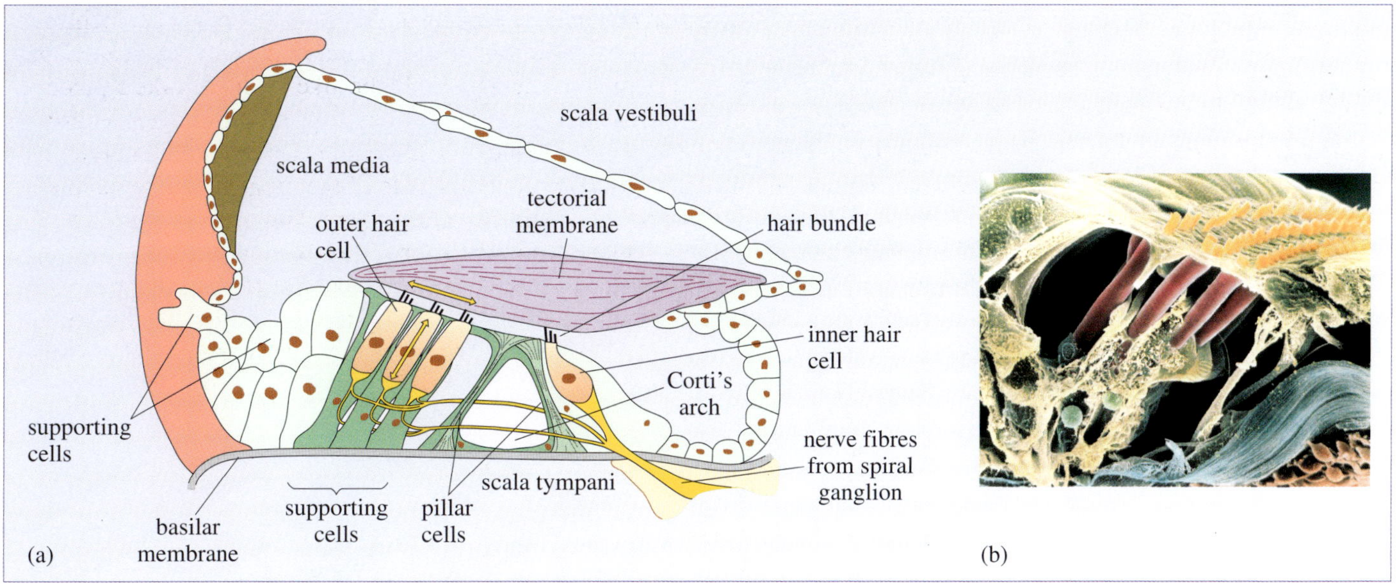

Figure 3.7 The organ of Corti: (a) sketch cross-section. (b) Coloured scanning electron micrograph. Four rows of hair cells can be seen, which are supported by pillar cells. (Magnification × 600.)

3.2.2 The role of the basilar membrane in sound reception

So far we know that sound-induced increases and decreases in air pressure move the tympanum inwards and outwards. The movement of the tympanum displaces the malleus which is fixed to its inner surface. The motion of the malleus and hence the incus results in the stapes functioning like a piston – alternately pushing into the oval window and then retracting from it. Since the oval window communicates with the scala vestibuli, the action of the stapes pushes and pulls cyclically on the fluid in the scala vestibuli. When the stapes pushes in on the oval window, the liquid in the scala vestibuli is displaced. If the membranes inside the cochlea were rigid, then the increase in fluid pressure at the oval window would displace the fluid up the scala vestibuli, through the helicotrema and down the scala tympani causing the round window to bulge out. This is actually a fairly accurate description of what happens except that the membranes inside the cochlea are not rigid. As a consequence, the increase in pressure in the cochlear fluid caused by the inward movement of the stapes also displaces fluid in the direction of the cochlear partition, which is deflected downwards. This downward deflection in turn causes the elastic basilar membrane to move down and also increases the pressure within the scala tympani. The enhanced pressure in the scala tympani displaces a fluid mass that contributes to outward bowing of the round window. When the stapes pulls back, the process is reversed and the basilar membrane moves up and the round window bows inwards. In other words, each cycle of a sound stimulus evokes a complete cycle of up-and-down movement of the basilar membrane and provides the first step in converting the vibration of the fluid within the cochlea into a neural code. The mechanical properties of the basilar membrane are the key to the cochlea's operation.

One critical feature of the basilar membrane is that it is not uniform. Instead, its mechanical properties vary continuously along its length in two ways. First, the membrane is wider at its apex compared to the base by a factor of about 5, and second, it decreases in stiffness from base to apex, the base being 100 times stiffer.

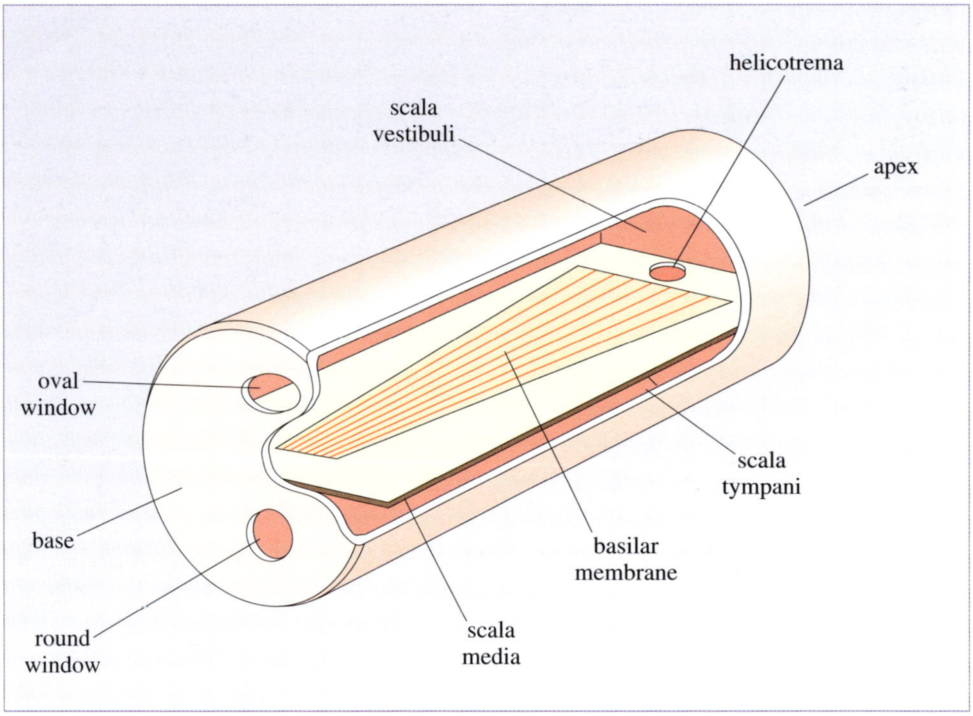

Figure 3.8 Schematic representation of the basilar membrane (cochlea uncoiled) showing the variation in width along its length.

So, the base is narrow and stiff compared to the apex (Figure 3.8). This means that stimulation by a pure tone results in a complex movement of the membrane. If it were uniform, then the fluctuating pressure difference between the scala vestibuli and the scala tympani caused by the sound would move the entire membrane up and down with similar excursions at all points. However, because of the variation in width and stiffness along its length, various parts of the membrane do not oscillate in phase. Over a complete cycle of sound each segment of the membrane undergoes a single cycle of vibration but at any point in time some parts of the membrane are moving upwards and some parts are moving downwards. The overall pattern of movement of the membrane is described as a **travelling wave**.

To visualize the motion of a travelling wave, think of a wave that travels along a piece of ribbon if you hold one end in your hand and give it a flick. Figure 3.9a is a representation of what you might expect by flicking a ribbon. Figure 3.9b represents a more realistic representation of the wave on the basilar membrane because the basilar membrane is attached at its edges and is displaced in response to sound in a transverse (crosswise) direction as well as a longitudinal direction.

- ❍ What do you notice about the change in amplitude of the wave as it travels along the membrane?
- ● As it travels, the wave reaches a peak amplitude that then rapidly falls. The amplitude of the wave is therefore greatest at a particular location on the membrane.

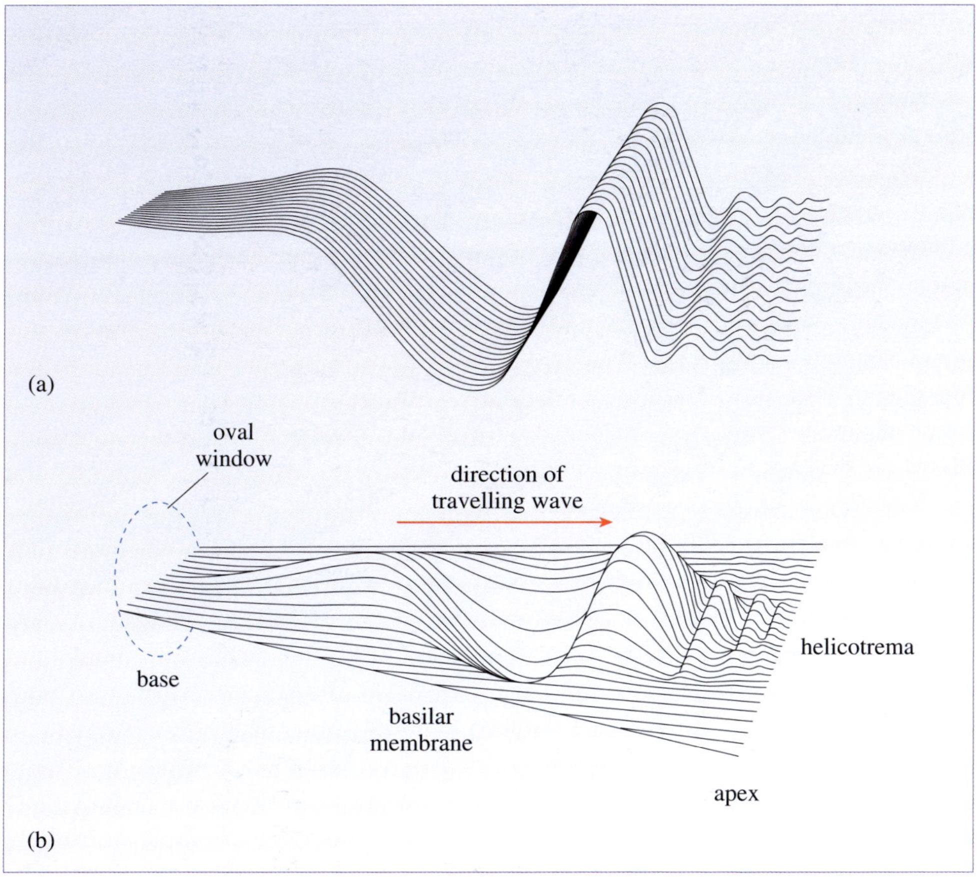

Figure 3.9 Instantaneous pattern of a travelling wave along the basilar membrane. (a) The pattern that would result if the membrane were ribbon-like. (b) The vibration of the membrane represented more realistically.

A travelling wave then, is a unique moving waveform whose point of maximal displacement traces out a specific set of locations. The shape described by the set of these locations along the basilar membrane is called the **envelope** of the travelling wave (Figure 3.10). The point along the basilar membrane where the wave, and hence the envelope traced by the travelling wave, reaches a peak differs for each frequency. In other words, each point along the basilar membrane that is set in motion vibrates at the same frequency as the sound impinging on the ear, but *different frequency* sounds cause a *peak* in the wave at *different positions* on the basilar membrane (Figure 3.11a overleaf).

Look at Figure 3.11b.

❍ What do you notice about the point of maximum displacement for each frequency?

● For the lowest frequency (60 Hz) the maximum displacement is near the apical end, for the highest frequency (2000 Hz) the maximum displacement is near the base, while the intermediate frequency has maximal displacement between the two.

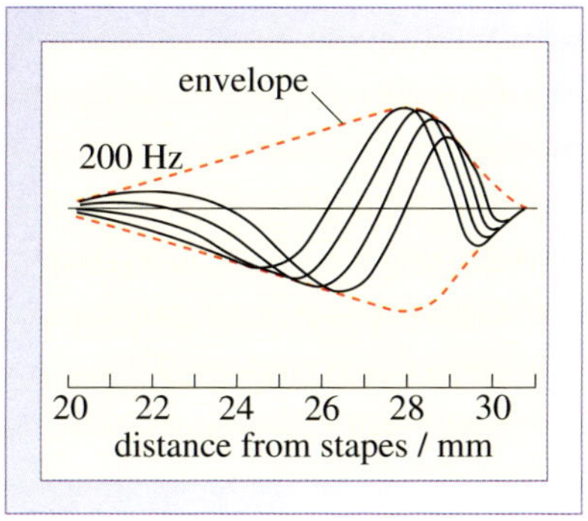

Figure 3.10 The envelope formed by a 200 Hz tone. The shape of the envelope is described by the set of momentary locations (four shown here) traced by the travelling wave along the basilar membrane.

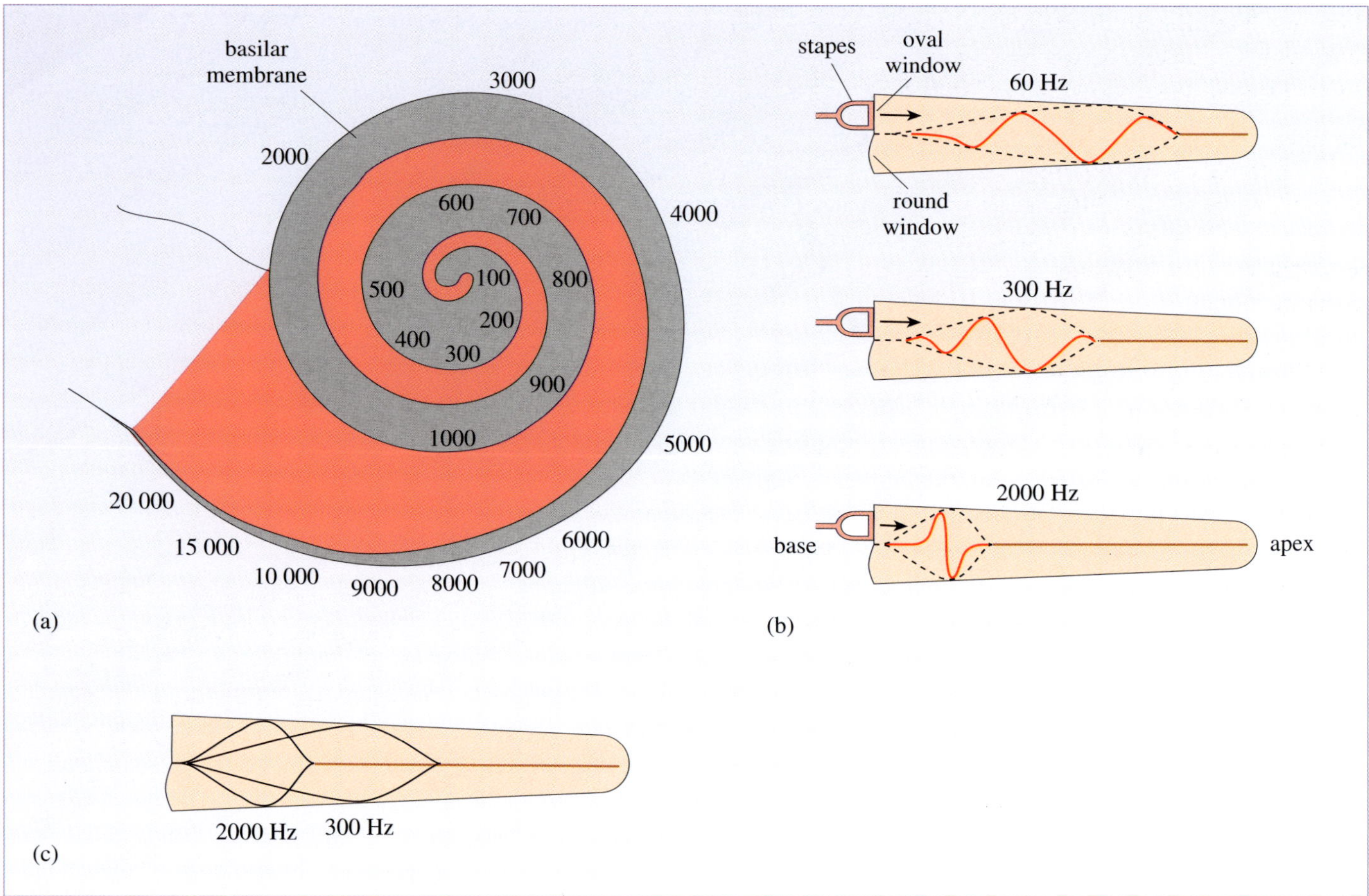

Figure 3.11 (a) A highly schematic map of frequency representation on the basilar membrane showing that the part of the basilar membrane that responds to sound depends on the frequency of the sound. (b) A schematic representation of the cochlea and the envelope of a travelling wave that would occur for stimuli of three different frequencies. One instantaneous waveform is shown for each frequency. (c) Displacement of the basilar membrane in response to a signal composed of two sinusoidal waves of 300 Hz and 2000 Hz.

Therefore, high-frequency sounds cause a small region of the basilar membrane near the stapes to move, while low frequencies cause almost the entire membrane to move. However, the *peak* displacement of the membrane is located near the apex. This shows that the travelling wave always travels from base to apex, and how far towards the apex it travels depends on the frequency of stimulation; lower frequencies travel further.

- ❍ What would the response of the membrane be if the sound impinging on the ear was a complex sound consisting of frequencies of 300 Hz and 2000 Hz?
- ● Each frequency would create a maximum displacement at a different point along the basilar membrane (as shown in Figure 3.11c).

The separation of a complex signal into two different points of maximal displacement along the membrane, corresponding to the sinusoidal waves of which the complex signal is composed, means that the basilar membrane is performing a type of spectral (Fourier) analysis much like that described in Section 2.3. The basilar membrane displacement therefore provides useful information about the frequency of the sound impinging on the ear by acting like a series of band-pass filters. Each section of the membrane passes, and therefore responds to, all sinusoidal waves with frequencies between two particular values. It does not respond to frequencies that are present in the sound but fall outside the range of frequencies of that section.

The filter characteristics of the basilar membrane can be studied using the technique of laser interferometry. Figure 3.12 shows the results of such a study. The data were collected by presenting different frequency sounds to the inner ear of a chinchilla and then measuring the level of each tone that is required to displace the basilar membrane by a fixed amount. Measurements are taken at a particular point on the basilar membrane.

❍ From Figure 3.12, determine the frequency of the tone that required the lowest sound level to displace the basilar membrane by a set amount.

● A little under 10 000 Hz (in fact 8350 Hz or 8.35 kHz).

This frequency is known as the **characteristic**, **critical** or **central frequency (CF)** of that part of the membrane because it is most sensitive to (or tuned to) frequencies in the region of 8 kHz.

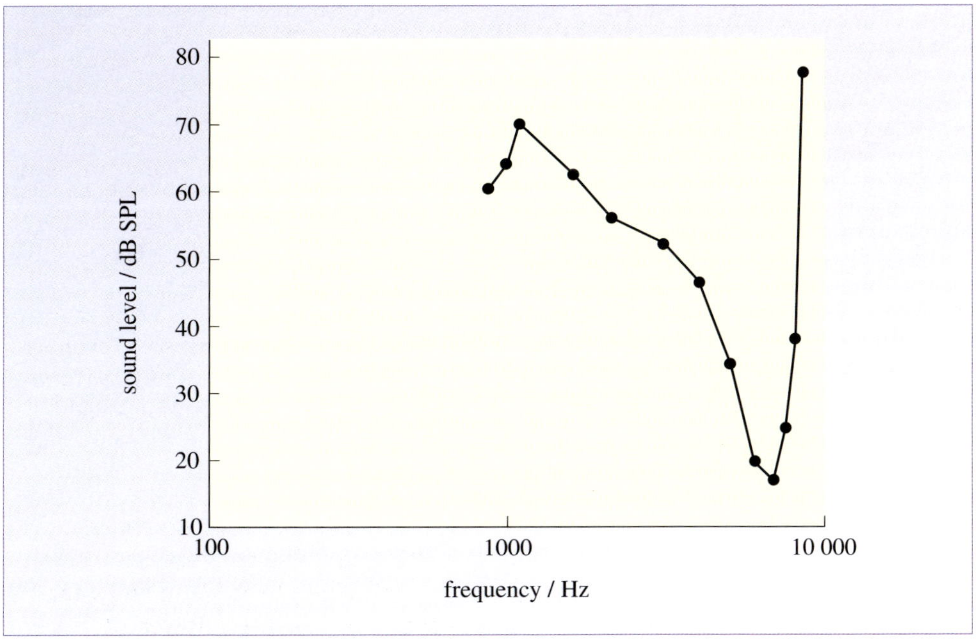

Figure 3.12 The sound level required to maintain the basilar membrane at a constant displacement (1.9×10^{-8} m) as a function of the frequency of the tonal input.

For frequencies above and below 8.35 kHz the tone had to be more intense in order to vibrate the membrane to the same extent as that caused by the 8.35 kHz tone. This particular point on the membrane therefore acts as a filter in that it responds maximally to tones of 8.35 kHz, but shows very little response to tones that are higher or lower than this.

In the next section we shall see how the band-pass filtering characteristics of the basilar membrane are preserved in the discharge pattern of nerve fibres that leave the cochlea.

The motion of the basilar membrane also provides information about the temporal pattern of acoustic stimulation: it takes longer for a low-frequency stimulus to reach its point of maximum displacement on the membrane than it does a high-frequency stimulus.

❍ Why is this?

● Because high-frequency stimuli cause maximal displacement of the membrane near the base of the cochlea (near the stapes), whereas low frequencies cause maximal displacement at the apical end. If the sound always travels from base to apex, it takes longer for the wave to travel to reach the apex.

Finally, the mechanics of the basilar membrane provide information regarding the level of acoustic stimulation. The greater the stimulus level, the greater the amount of basilar membrane displacement. Therefore, more intense signals cause greater membrane displacement at a particular point than less intense stimuli.

You should now read Chapter 1 of the Reader, *The mechanics of hearing* by Jonathan Ashmore. There may be some terms and concepts that will not be familiar to you. Do not worry too much at this stage. There is some overlap in the material covered in the chapters and some of the concepts mentioned in Chapter 1 will be more comprehensively covered in Chapter 2 and later sections of the block.

In general, you may find it useful initially to read the chapters of the Reader for this block at the points indicated in the block text, with the main aim of getting a feel for what is covered in each chapter. Then, once you have completed the block, read the chapters through more carefully.

Summary of Sections 3.0 to 3.2.2

The ear is made up of the outer, middle and inner ears. The outer ear consists of the pinna, the external auditory canal and the tympanic membrane. The middle ear is air-filled and contains the middle ear ossicles. The inner ear is fluid-filled and contains the cochlea, the semicircular canals and the vestibule.

Sound in the external environment is channelled into the auditory meatus by the pinna and impinges on the tympanic membrane causing it to vibrate. These vibrations are transmitted to the inner ear via the middle ear ossicles. The ossicles act as an impedance-matching device and amplify vibrations between the outer and inner ear. They also function in preventing damage to the inner ear by very loud sounds via the middle ear reflex.

The inner ear contains the cochlea which has three compartments: the scala tympani, the scala vestibuli and the scala media (cochlear duct). Inside the cochlear duct is the organ of Corti. The organ of Corti contains the sensory receptors that are called hair cells, sits on top of the basilar membrane and is covered by the tectorial membrane.

The stapes connects to the scala vestibuli via the oval window. Movement of the stapes in response to sound causes the fluid in the scala vestibuli to vibrate. This causes the basilar membrane to move. The motion is described as a travelling wave. The base of the membrane is 5 times narrower and about 100 times stiffer than the apex.

The basilar membrane has a frequency-to-place conversion for pure-tone stimuli. High-frequency sounds cause greatest vibration near the base of the membrane, and low frequencies cause greatest vibration near the apex.

The basilar membrane acts like a band-pass filter. Each point on the membrane corresponds to a band-pass filter with a different centre frequency. This means that sounds of different frequency result in maximal displacement at different points along the membrane.

3.2.3 The organ of Corti and hair cells

We have established that the vibration patterns of the basilar membrane carry information about frequency, amplitude and time. The next step is to examine how this information is converted or coded into neural signals in the auditory nervous system. To do so, we must look at the organ of Corti in some detail since it is here that the auditory receptor cells that convert mechanical energy into a change in membrane polarization are located.

As we saw in Section 3.2.1, the receptor cells responsible for the transduction of mechanical energy into neural energy are called hair cells. A typical hair cell is shown in Figure 3.13. The outer hair cells are closest to the outside of the cochlea and are arranged in 3 rows whereas the inner hair cells form a single row (see Figure 3.7). There are about 12 000 outer hair cells and about 3500 inner hair cells in the human ear. The tips of the tallest row of cilia of each hair cell are in contact with the tectorial membrane which is situated at the top of the organ of Corti and has a soft, ribbon-like structure.

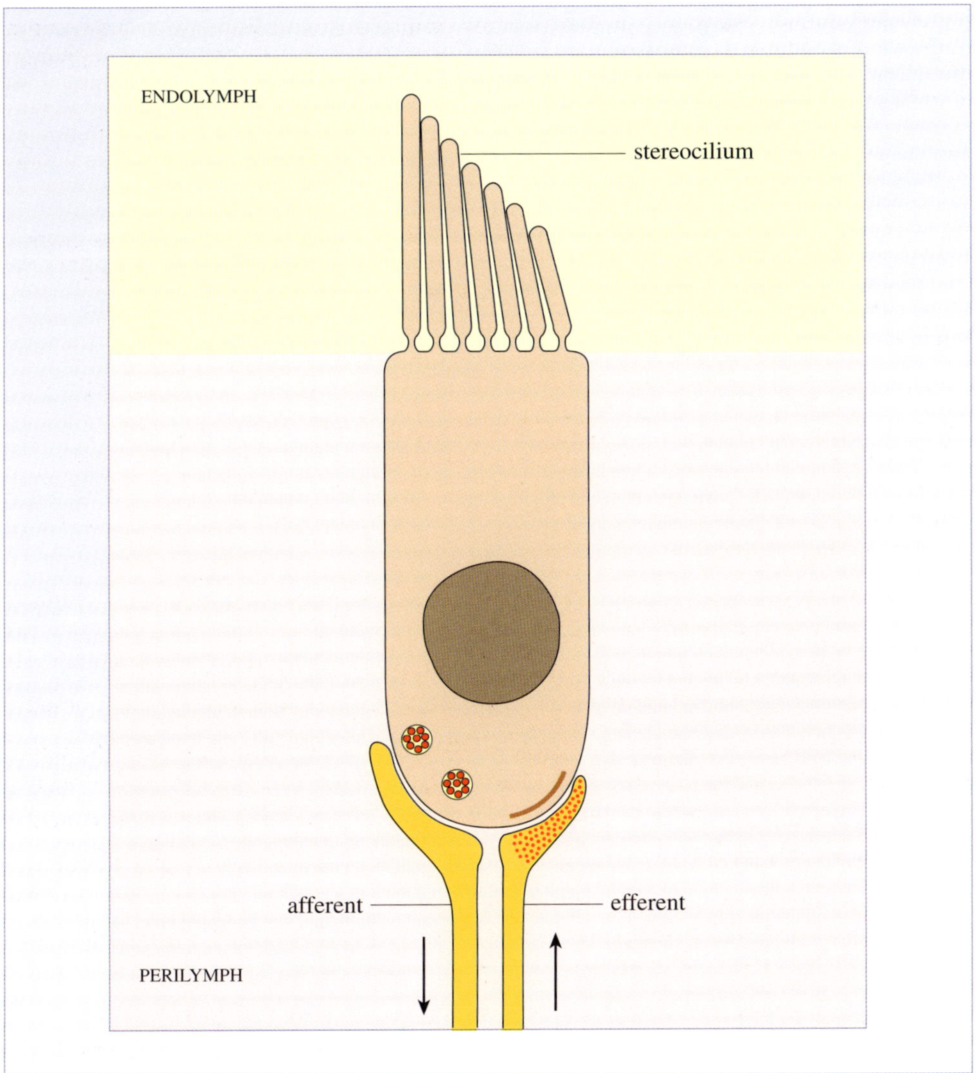

Figure 3.13 The structure of a hair cell. Inside the cell body can be seen the nucleus, a synaptic bar, and synaptic vesicles which carry the neurotransmitter molecules.

3.2.4 Neural transduction

The critical event for the transduction of sound into a neural signal is the bending of the stereocilia of the hair cells. In this section we will examine how the flexing of the basilar membrane leads to the bending of the stereocilia and the production of a neural signal.

I Hair cells transform mechanical energy into neural signals

The tectorial membrane runs parallel to the basilar membrane, so when the basilar membrane vibrates up and down in response to motion at the stapes, so does the tectorial membrane. However, as shown in Figure 3.14, the displacement of the membranes causes them to pivot about *different* hinging points and this creates a shearing force between the hair cell stereocilia embedded in the tectorial membrane and the hair cells themselves which rest on the basilar membrane. Shearing is a particular form of bending in which, in this case, the top moves more than the bottom. It is this shearing force that transduces mechanical energy into electrical energy which is transmitted to the auditory nerve fibres.

❍ What kind of sensory receptor transduces mechanical energy into electrical energy?

● Mechanoreceptors.

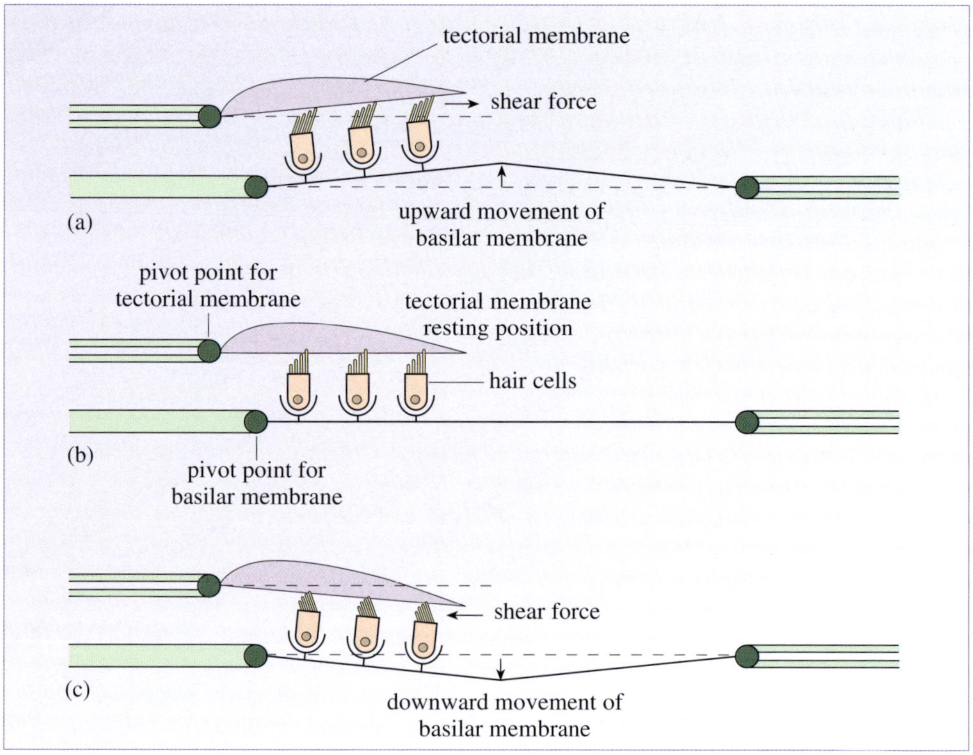

Figure 3.14 Schematic diagrams of shearing forces created between the hair cells and the tectorial membrane as a result of basilar membrane displacement. (a) Shearing force that results from displacement of the basilar membrane towards the scala vestibuli when the basilar membrane is driven upwards. (b) Relationship between hair cells and tectorial membrane with no stimulation. (c) Shearing forces in the direction opposite to that shown in (a) after displacement in the opposite direction.

In order for the hair cell to transduce stereocilia shearing (mechanical) forces into an electrical (neural) response, the permeability of the hair cell membrane must change. This happens when the shearing motion, which is a mechanical stimulus, opens **ion channels** in the cell's plasma membrane and the current flowing through these channels alters the cell's membrane potential (this is the electrical response). So, in response to a mechanical stimulus, there is an influx of ions into the cell which disturbs the resting potential of the cell membrane, driving the membrane potential to a new level called the **receptor potential**. The channels are relatively non-selective about which ions they allow to pass through them. However, you should recall from Section 3.2.1 and from Chapter 1 of the Reader, that potassium is very plentiful in the endolymph. The stereocilia of the hair cells are bathed in endolymph whereas the basal region of the cell is bathed in perilymph (which is relatively low in potassium). So once the channels are opened, potassium ions flow into the hair cell.

❍ How does this differ from most other cells?

● In most cells, the flow of potassium ions is outwards because the cell is higher in potassium than the surrounding medium. In the case of hair cells however, the endolymph surrounding the stereocilia has a higher level of potassium than the cell, and the flow is in the reverse direction.

In fact, when a hair bundle is displaced by a mechanical stimulus, its response depends on the direction and magnitude of the stimulus. In an unstimulated cell about 10% of the ion channels are open. As a result, the cell's resting potential (about −50 mV) is determined, in part, by the inward flow of current. A positive stimulus that displaces the stereocilia towards the tall edge opens additional channels and the resultant influx of positive ions depolarizes the cell by as much as tens of mV. A negative stimulus that displaces the stereocilia towards the short edge shuts the channels that are open at rest and hyperpolarizes the cell (Figure 3.15 overleaf).

This directional sensitivity of the cells, their arrangement on the organ of Corti and the hypothesized motion of the organ of Corti in response to a stimulus, means that an upward movement of the basilar membrane leads to depolarization of the cells, whereas a downward deflection elicits hyperpolarization.

The receptor potential of a hair cell is graded; as the stimulus amplitude increases, the receptor potential grows increasingly larger, up to a maximal point of saturation. The relationship between a bundle's deflection and the resulting electrical response is S-shaped (Figure 3.15d). This results in a high degree of sensitivity. A small displacement of only 100 nm (100×10^{-9} m) represents 90% of the response range of the hair cell (shaded part). Deflection of a hair cell by the width of a hydrogen atom is enough to make the cell respond.

2 *Mechanical force directly opens and closes transduction channels*

It is believed that **tip links** aid in causing 'channels' to open and close near the top of the hair cell (Figure 3.16 overleaf). Tip links are filamentous connections between two stereocilia. Each tip link is a fine fibre obliquely joining the distal end of one stereocilium to the side of the longest adjacent process. It is thought that each link is attached at one end or both to the molecular gates of one or a few channels. Under this arrangement, pushing a bundle in one direction increases the tension on the tip link and promotes channel opening while pushing the bundle in the opposite direction slackens the link and the associated channel closes.

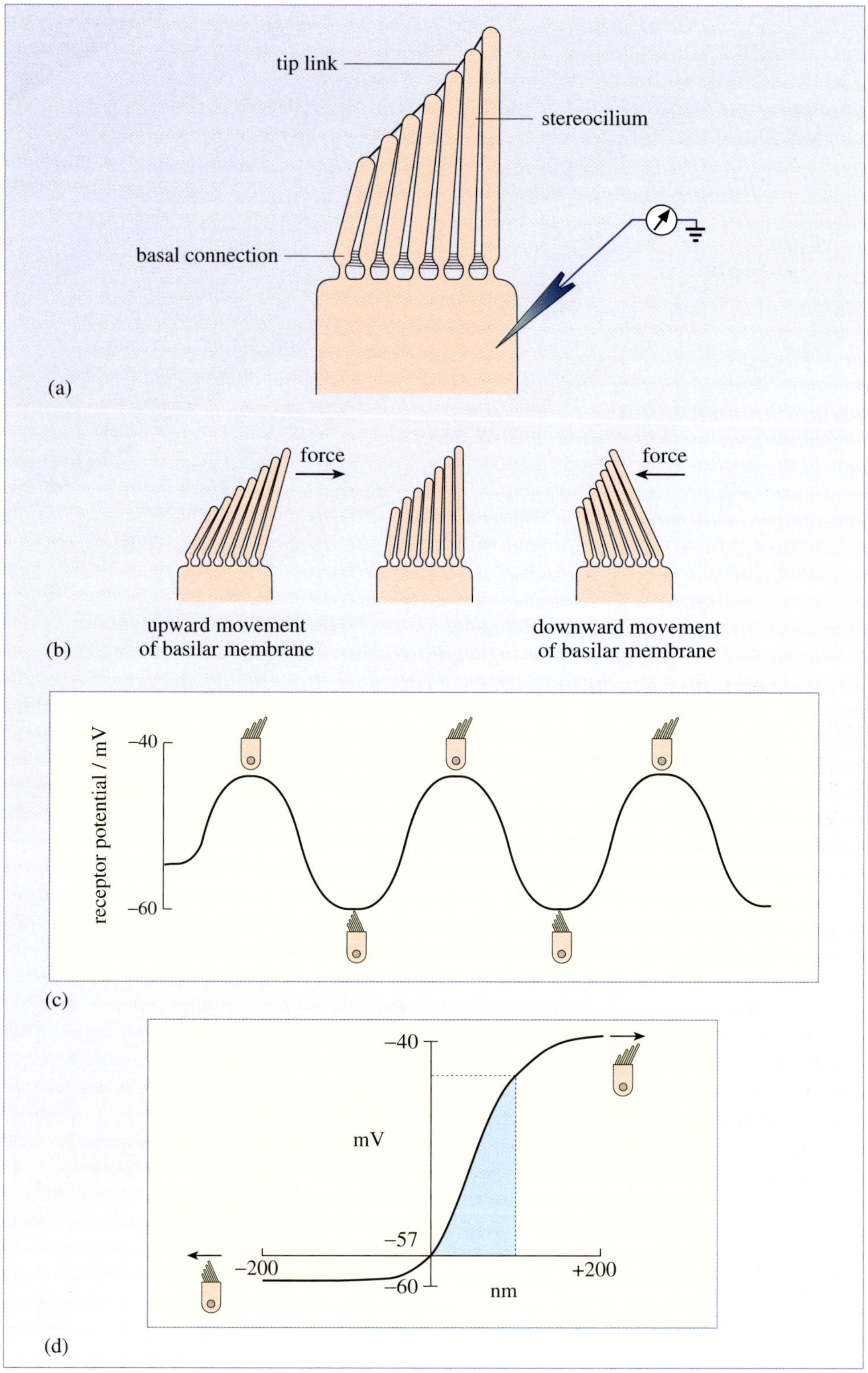

Figure 3.15 Sensitivity of hair cells. (a) A schematic drawing of a hair cell with an electrode inserted into its cytoplasm. (b) Application of a mechanical force to the hair bundle causes a deflection in the stereocilia. (c) When the top of a hair bundle is displaced back and forth by a stimulus probe, the opening and closing of mechanically-sensitive channels produces an oscillatory receptor potential. (d) The sigmoidal relationship between hair bundle deflection (horizontal axis) and receptor potential (vertical axis) in a stimulated hair cell. The shaded area shows that a small displacement of only 100 nm represents 90% of the response range of the hair cell.

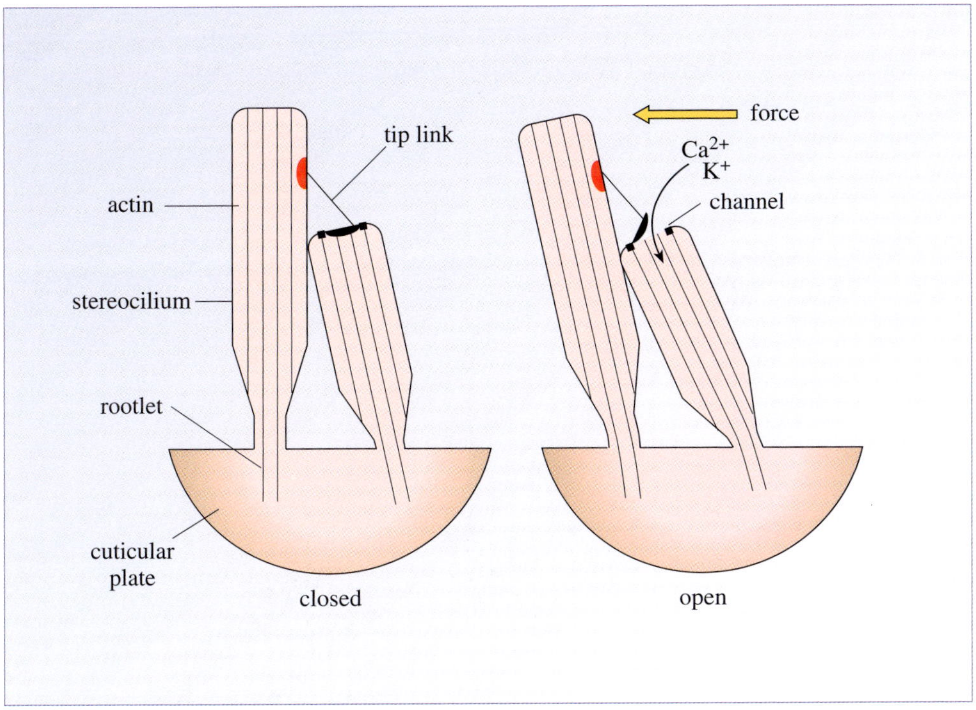

Figure 3.16 Mechanism of mechano-electrical transduction. When the hair bundle is at rest each transduction channel oscillates between closed and open states, spending most of the time closed. Displacement of the bundle in the positive direction increases tension on the tip link and promotes channel opening, the influx of cations and a depolarizing receptor potential.

3 *Mechano-electrical transduction is rapid*

Many other sensory receptors, such as photoreceptors and olfactory neurons, employ second messengers in the transduction process. This is not true for hair cells. The rapidity with which they respond makes this impossible. In order to deal with the frequencies of biologically relevant stimuli, transduction must be rapid. The highest frequency humans can hear is about 20 000 Hz. This in effect means that hair cells must be able to turn current on and off 20 000 times per second (200 000 times per second for a bat). Also, localization of sound sources (Section 5.4) requires that animals are able to resolve very small time differences, in the order of 10 μs.

3.2.5 Synaptic transmission from hair cells

In addition to being sensory receptors, hair cells are also presynaptic terminals. The membrane at the base of each hair cell contains several presynaptic active zones, where chemical neurotransmitter is released. When the hair cells are depolarized, chemical transmitter is released from the hair cells to the cells of the auditory nerve fibres. Excited by this chemical transmitter, the afferent nerve fibres contacting the hair cells fire a pattern of action potentials that encode features of the stimulus. We will return to how this information is encoded in Section 4.2. As in other synapses, the depolarization that leads to transmitter release acts through an intermediary, namely calcium ions. Depolarization opens channels at the base of the hair cell (voltage-gated calcium channels), which allow calcium ions to enter from the surrounding perilymph resulting in the release of transmitter (Figure 3.17 overleaf). Calcium also has another function: it opens potassium channels, called calcium-activated potassium channels, which allow potassium ions to leave the cells because the perilymph on the other side is low in potassium. The potassium ions leaving the hair cell via the calcium-activated channels results in repolarization of the cell. The identity of the neurotransmitter is controversial. Glutamate appears to be the transmitter in some cases but there is also evidence for another, as yet unidentified, substance.

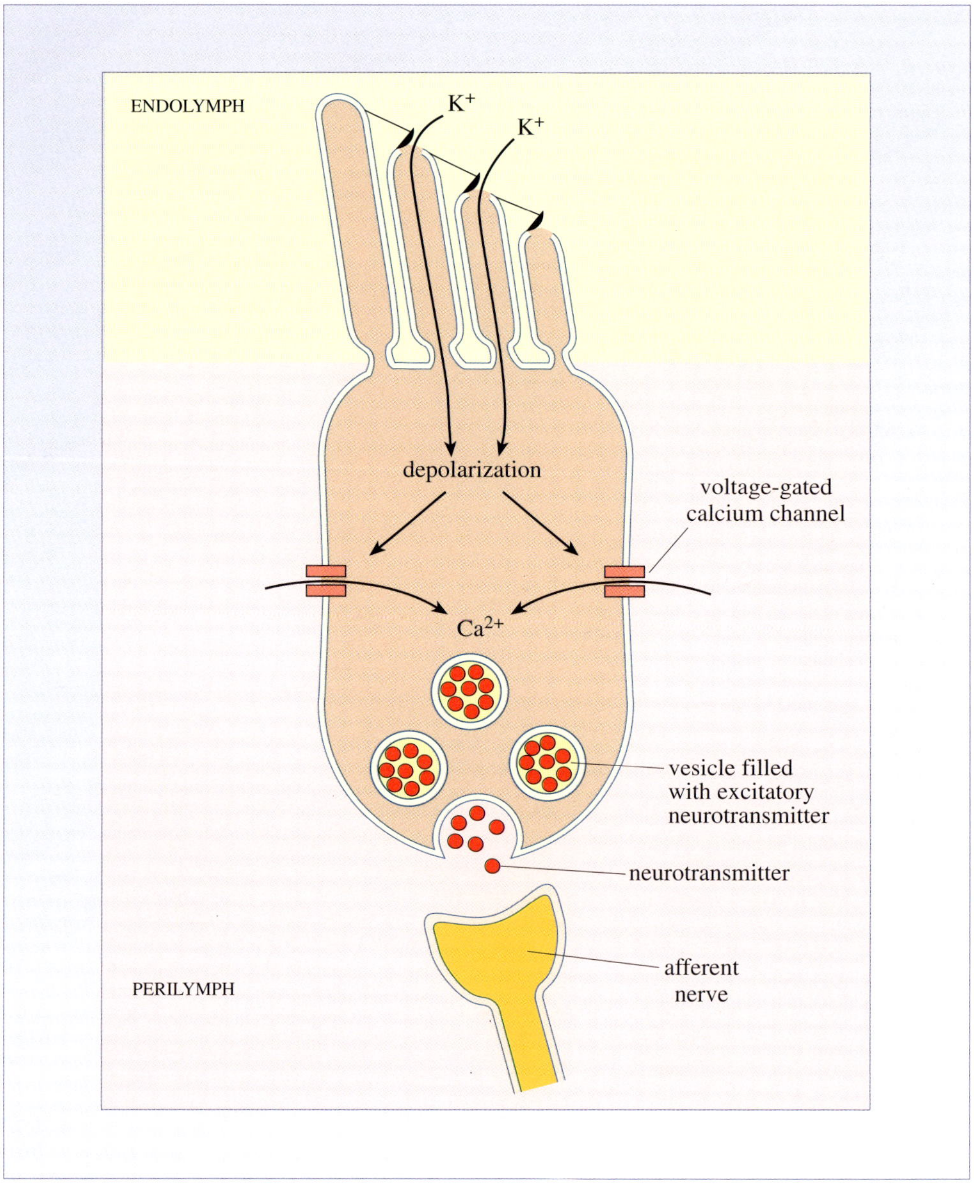

Figure 3.17 Depolarization of a hair cell. Entry of potassium ions depolarizes the hair cell which opens voltage-gated calcium channels. Incoming calcium ions further depolarize the cell leading to the release of chemical transmitter to the afferent nerve fibre contacting the hair cell.

3.2.6 Hair cell tuning

We have determined that the location of the peak of the travelling wave on the basilar membrane is determined by the frequency of the originating sound. The hair cells run the length of the basilar membrane. When a certain frequency sound stimulates a point on the membrane, it responds by moving, and the hair cells at that site are stimulated by the shearing force that this movement creates (Figure 3.18). Groups of hair cells therefore only respond if certain frequencies are present in the originating sound. The frequency sensitivity of a hair cell can be displayed as a **tuning curve**. To construct a tuning curve, a single hair cell is stimulated repeatedly with pure tone stimuli of various frequencies. For each frequency, the intensity of the stimulus is adjusted until the response of the hair cell reaches some predefined level. The tuning curve is then the graph of sound intensity against stimulus frequency (Figure 3.19). Tuning curves for hair cells are characteristically V-shaped. The tip represents the frequency to which the cell is most sensitive.

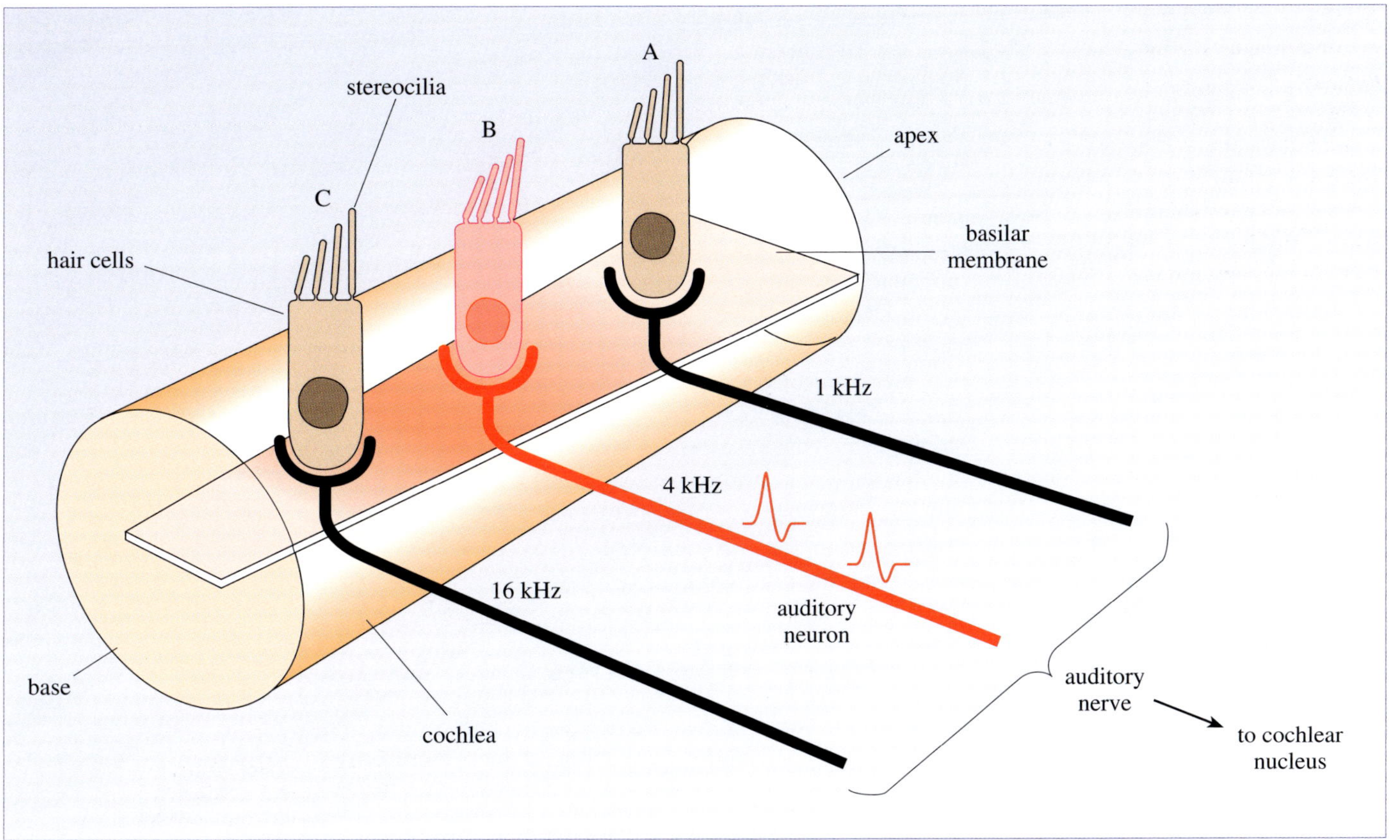

▲ **Figure 3.18** Schematic diagram of the basilar membrane and hair cell tuning. A 4 kHz sound results in a peak in the travelling wave at position B. The hair cell at this position is stimulated by the bending of the stereocilia. The depolarization results in transmitter release and the generation of an action potential in the auditory nerve fibre.

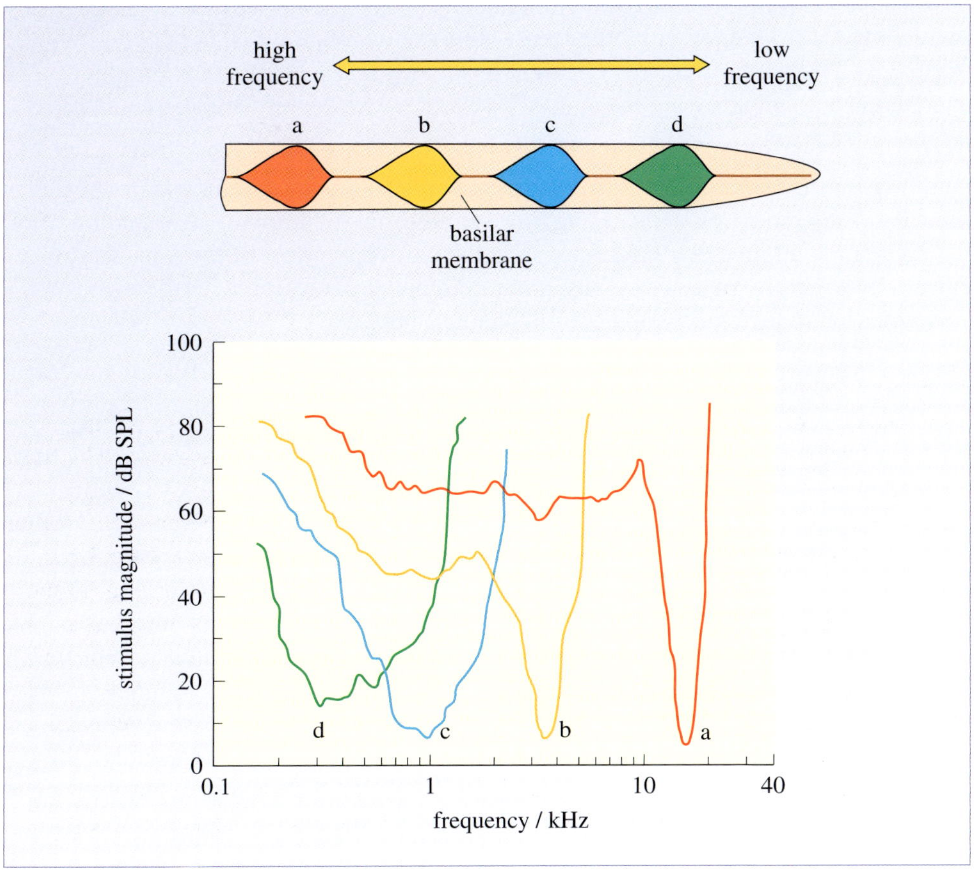

Figure 3.19 Tuning curves of hair cells located at different positions on the basilar membrane.

A sound of this frequency will elicit a response from the cell even when it is of very low intensity. Sounds of greater or lesser frequency require higher intensity to excite the cell to the predetermined level.

Adjacent piano strings are tuned to frequencies some 6% apart. On average, successive inner hair cells differ in characteristic frequency by about 0.2%.

The great majority of neurons that carry information from the cochlea to higher levels of the auditory system connect to the inner hair cells. Thus most, if not all, information about sounds is conveyed to the brain via the inner hair cells. Given that the outer hair cells greatly outnumber the inner hair cells, it seems paradoxical that most cochlear output is derived from the inner cells. However, ongoing research suggests that outer hair cells do play an important role in the transduction process. Membranes of the outer cells contain a **motor protein** that changes the length of the outer hair cells in response to stimulation. This change in length effects a change in the mechanical coupling between the basilar and tectorial membranes. Outer hair cells are sometimes said to constitute a **cochlear amplifier** by amplifying the response of the basilar membrane. This causes the sterocilia on the inner cells to bend more, creating a bigger response in the auditory nerve (Figure 3.20).

Summary of Sections 3.2.3 to 3.2.6

Hair cells are found in the organ of Corti and run the length of the basilar membrane. They transform mechanical energy into neural signals.

When the basilar membrane vibrates in response to sound, hair cells located at the site of maximal vibration on the basilar membrane are stimulated. This means that the mechanical properties of the membrane allow the auditory system to distinguish one frequency from another by the location on the membrane that is maximally excited by a particular frequency. Hair cells located at the place of maximum excitation respond, allowing the auditory system to extract information about the frequency of a sound.

Auditory transduction occurs when the basilar membrane moves up and down and the cilia of the inner hair cells rub against the tectorial membrane. The bending of the cilia produces an electrical response in the hair cells. Most of the transduction current is carried by potassium ions, potassium being the cation with the highest concentration in the endolymph bathing the hair bundle.

Displacement of the stereocilia towards the tall edge results in an influx of cations and a depolarization of the hair cell. Displacement of stereocilia towards the short edge results in hyperpolarization of the hair cell. Depolarization of the hair cell allows calcium ions to enter the cell leading to the release of transmitter from the presynaptic terminals on the hair cell.

Outer hair cells help amplify vibrations of the basilar membrane.

Question 3.1

Discuss the two ways in which the middle ear increases the effectiveness with which sound is transmitted from the external ear to the inner ear.

Question 3.2

Draw a flow-diagram to illustrate the route a sound (pressure) wave takes from the time it enters the external ear to the point at which it reaches the round window.

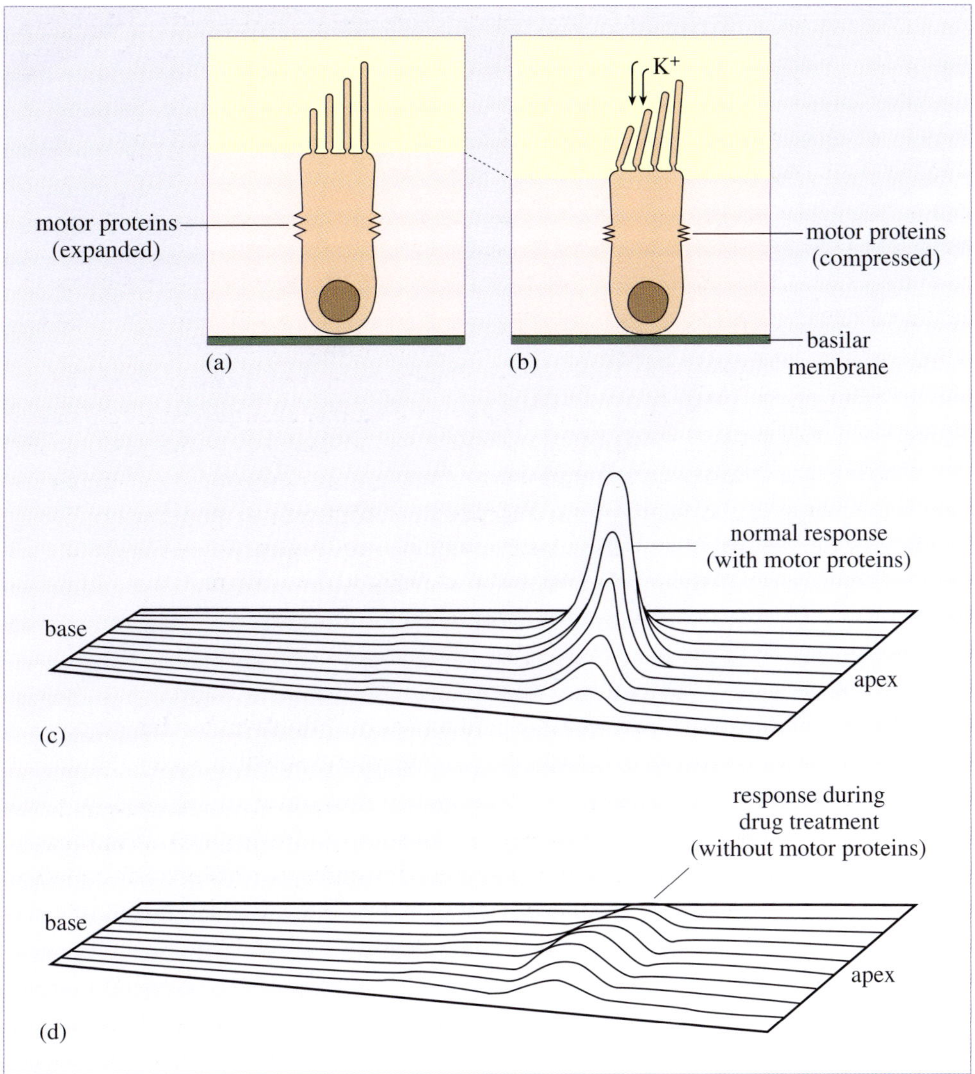

Figure 3.20 Amplification by outer hair cells. (a) Motor proteins in the membranes of the outer hair cells are expanded when the cells are in a resting state. (b) When potassium enters the cell, motor proteins are activated and contract the hair cell. (c) Conformational change in the hair cell increases the bending of the basilar membrane. (d) If the cochlear amplifier is deactivated (for example, with drugs) bending of the basilar membrane is decreased dramatically (see Reader Chapter 2, Figure10).

Question 3.3

Describe the basic structure of the cochlea and discuss how the different structures contribute to the reception of sound.

Question 3.4

What is a travelling wave in the context of the response of the basilar membrane to an incoming sound signal?

Question 3.5

What are the different properties of the fluids found in the main compartments of the cochlea? How do they contribute to the transduction of a neural signal?

Neural processing of auditory information

4

In this section we will look at how the frequency selectivity found along the basilar membrane is preserved or modified by the auditory nerve and how information about the intensity of the signal is encoded in the response of the auditory nerve fibres.

The nerve that communicates with or innervates the hair cells along the basilar membrane is called the **vestibulocochlear nerve** or VIIIth cranial nerve. It enters the brainstem just under the cerebellum and conveys information from the hair cells in the inner ear as well as from the vestibular organs of the inner ear. The cochlear portion of the nerve (**auditory nerve**) contains two basic types of auditory nerve fibres: **afferent fibres** that carry information from the peripheral sense organ (organ of Corti) to the brain; and **efferent fibres** that bring information from the cerebral cortex to the periphery. Afferent fibres arise from nerve cell bodies in the **spiral** (or **cochlear**) **ganglion** (Figure 4.1) and contact the hair cells. The hair cells themselves do not have axons and therefore do not generate action potentials. Action potentials are first produced by the axons of afferent fibres. Recall that about 10% of the ion channels are open when the hair cell is unstimulated (Section 3.2.4). This means that in the auditory nerve, there is a continuous low level of discharge of action potentials even when hair cells are unstimulated. Depolarization of hair cells in response to stereocilia shearing causes an increase in the discharge rate of action potentials above this spontaneous rate (excitation) while hyperpolarization of hair cells leads to a decrease in the discharge rate of action potentials below the spontaneous discharge rate (inhibition) (Figure 4.2 overleaf).

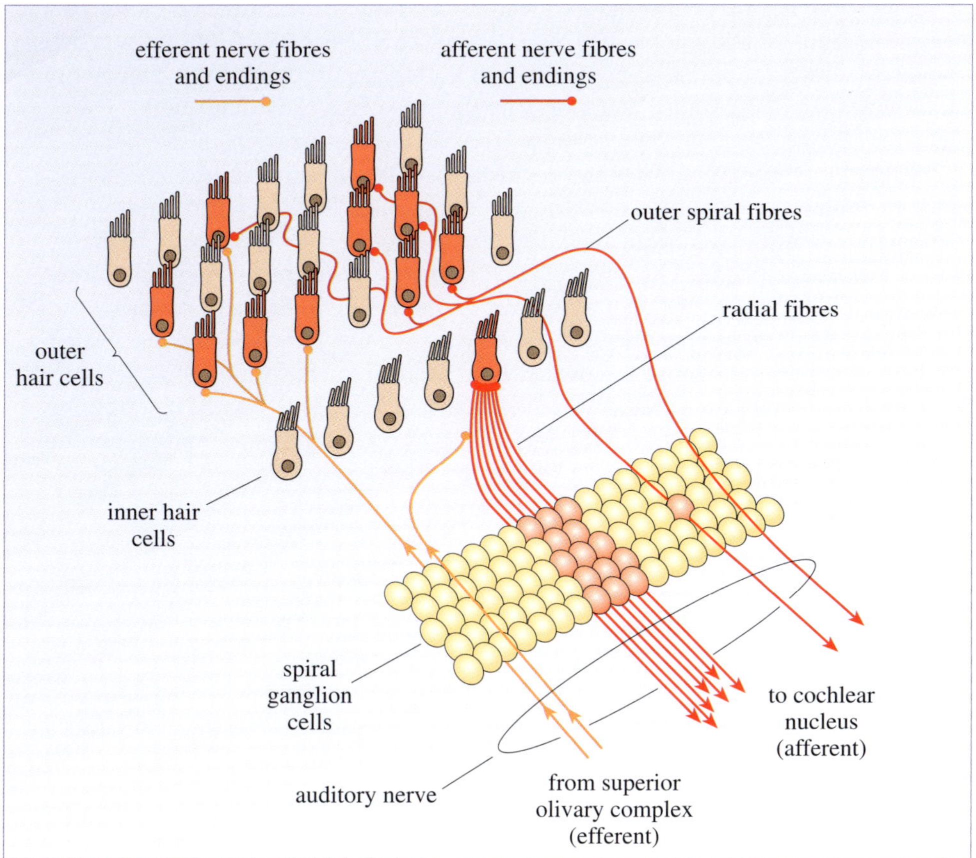

Figure 4.1 Innervation of the organ of Corti. Afferent fibres arise from nerve cell bodies within the spiral ganglion. Ninety-five per cent of afferents contact inner hair cells, each of which consists of the sole terminus for up to ten axons. Five per cent of afferents contact the outer hair cells.

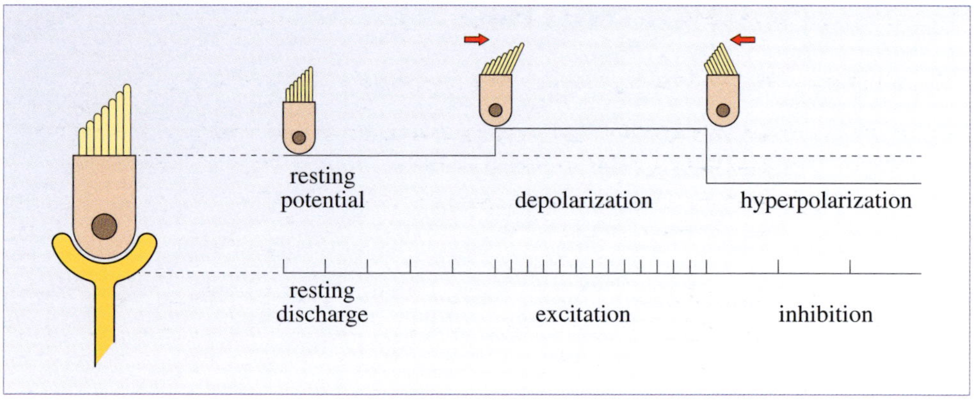

Figure 4.2 In unstimulated hair cells there is a low level of discharge of action potentials in the axons of the auditory nerve fibres. When the cell is stimulated, depolarization results in an increase in the discharge rate of action potentials (excitation) while hyperpolarization results in a decrease in discharge rate of action potentials (inhibition).

The inner hair cells are innervated by 95% of the afferent fibres. In humans an average of 8 fibres innervate 1 inner hair cell. They therefore make a many-to-one connection with inner hair cells (Figure 4.1). The other 5% of afferents innervate the outer hair cells.

4.1 Frequency coding in cochlear nerve fibres

Place code

We know that each hair cell occurs in a localized region of the cochlea, and that auditory nerve fibres contacting each hair cell fire action potentials in response to movement of the basilar membrane at that location. This means that the response of any given fibre should reflect the frequency selectivity of that location on the basilar membrane from which it comes. In other words, cochlear nerve fibres preserve the frequency selectivity found along the basilar membrane. Fibres on the outside of the auditory nerve bundle (those that innervate the basal hair cells) have high characteristic frequencies whereas those towards the middle of the nerve bundle (those that innervate the apex of the cochlea) have low characteristic frequencies. Thus, each place or location within the nerve responds 'best' to a particular frequency. The nerve fibres are spatially arranged to correspond directly to their basilar membrane origin. This arrangement is known as **tonotopic organization**, which can be defined as *the orderly spatial arrangement of neural elements corresponding to the separation of different frequencies*. Functionally, tonotopic organization allows the input frequency to be determined according to which nerve fibre discharges with the greatest relative discharge rate. This way of determining frequency is known as the **place theory** and gives rise to the **place code**. Tonotopic organization is found at all higher levels of the auditory system up to and including the auditory cortex.

There are several ways in which the frequency selectivity of single fibres can be determined. One way is to present a single fibre with a wide range of stimuli of different frequencies but identical intensity. The function generated when responses to the stimuli are plotted against the frequency of each stimulus is called an **iso-intensity contour**. Figure 4.3a shows a number of such contours for a single fibre in the auditory nerve. Each curve is generated using a different intensity level of the stimulus.

❍ What do you notice about the different contours?

- The higher the intensity of the stimulus, the broader the contour.

❍ What does this indicate in terms of the frequency selectivity of the fibre?

- It means that at lower intensities, the fibre responds maximally to a narrow range of frequencies but as the intensity of the signal increases, the range of frequencies to which the fibre responds increases, i.e. the fibre shows lower frequency selectivity at high intensities.

A second way of displaying the tuning characteristics of single auditory fibres is to generate a tuning curve. This is done the same way as tuning curves for hair cells are generated (Section 3.2.6). Figure 4.3b is a tuning curve for a hair cell from a guinea pig showing how the threshold intensity for a given fibre varies with stimulus frequency.

❍ What is the characteristic frequency (CF) of this particular fibre?

- About 18 kHz. The intensity of the stimulus needed to elicit a response is lower than for any other frequency (less than 40 dB SPL).

You can see that the high-frequency side of the curve is very steep whereas the low-frequency side is less steep and may have a long, low-frequency tail.

❍ What does this indicate?

- It means that nerve fibres are unlikely to respond to many frequencies higher than their characteristic frequency (CF) even when the intensity of the signal is very high. (At a frequency of about 23 kHz, the intensity of the signal needed to be about 70 dB to elicit a response.) For frequencies below CF the intensity of the signal needed to elicit a response is not quite so high and there is a broader range of frequencies to which the fibre will respond.

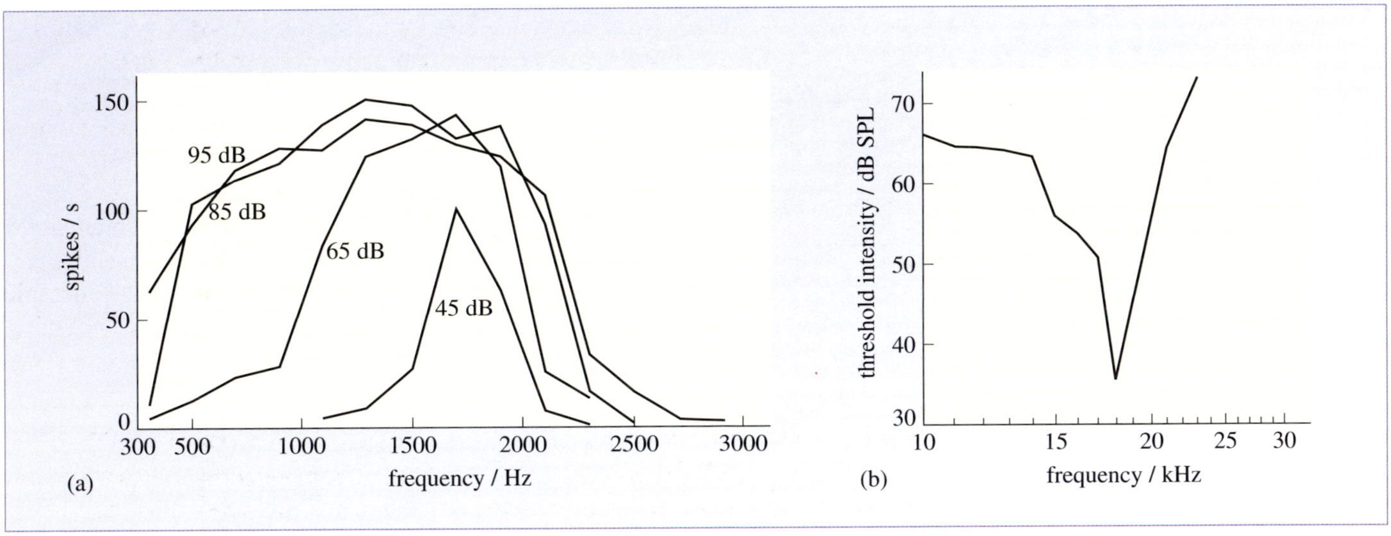

Figure 4.3 (a) Iso-intensity contours for a single fibre in the auditory nerve. For each contour generated, the intensity of the stimulus is kept constant while the frequency is varied. The number of spikes (action potentials) generated by the nerve fibre is then recorded for each frequency. The more spikes generated, the more sensitive or tuned the fibre is to that particular frequency. (b) Tuning curve for an auditory nerve fibre from a guinea pig.

Frequency code

Although the evidence for the place theory of frequency coding is compelling, there is some question as to whether the tuning curves obtained from neurons in the auditory nerve provide a mechanism for frequency discrimination that is fine enough to account for behavioural data. People can detect remarkably small differences in frequency – in some cases as small as 3 Hz (for a 1000 Hz signal at moderate intensity). What accounts for this ability? As early as 1930, the American experimental psychologists Glen Wever and Charles Bray proposed that in response to a pure tone, the vibration of the basilar membrane matches the input frequency. They further suggested that the auditory receptors respond in such a way that the temporal pattern of basilar membrane vibration is reproduced in the firing pattern of the neuron. This could be achieved if auditory nerve fibres respond by firing one or more action potentials at the same time in every cycle of a pure tone. This is known as a **phase-locked** response, since the response appears locked to a certain point (e.g. the peak) in the stimulus (Figure 4.4a). By phase locking, the response pattern of the nerve fibre would accurately reflect the frequency of the sound wave. This is called the **frequency code**.

This idea is attractive and there is evidence that it occurs, but only at low frequencies. The reason for this is that neurons cannot fire much faster than about

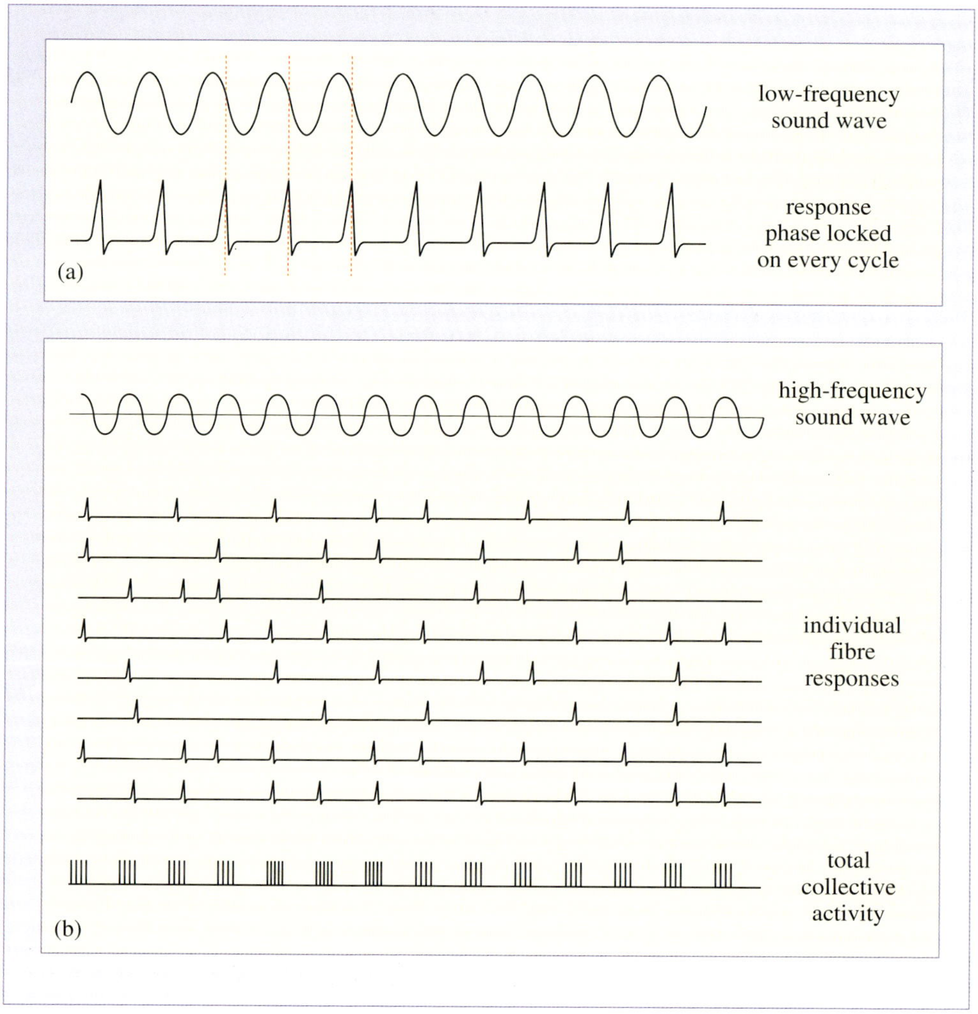

Figure 4.4 Phase locking. (a) The neuron is phase locked to the same point in every cycle of the pure tone stimulus. (b) The volley principle. The ensemble of fibre responses shown at the bottom of the figure has a pattern of firing that corresponds to the frequency of the input sound wave.

1000 action potentials per second (they have an **absolute refractory period** of about 1 ms and cannot fire twice in succession at intervals of less than 1 ms). Therefore they cannot fire during each cycle of the stimulus for stimuli above 1000 Hz (1 kHz). This realization led Wever and Bray to propose the operation of a **volley principle** illustrated in Figure 4.4b. In this figure, the frequency of the sound wave illustrated on top is too high for a single fibre to fire on every cycle. According to the volley principle each fibre only fires at a certain point in the cycle although it does not respond to *each* cycle. Each of the eight fibres illustrated is firing in phase; that is, if on any cycle a given auditory nerve fibre does fire, it does so at the same relative position within the cycle. If the responses of all fibres are then combined, as may happen further up in the auditory system, then information regarding signal frequency is preserved. The bottom trace in the figure shows the combined responses of all eight fibres; while none of the individual fibres reproduces the pattern of the wave, the combined response is sufficient to reproduce the frequency of the incoming signal. Using this principle, fibres can phase lock to signals with frequencies up to about 4 kHz thereby enhancing the transmission of information about stimulus frequency. Above this level, the variability inherent in neural firing becomes too great for such fine patterns to be resolved, and the frequency is probably coded for solely by the place code.

The central nervous system therefore gains information about stimulus frequency in two ways. First there is the place code: the fibres are arranged in a tonotopic map such that position is related to characteristic frequency. Second, there is the frequency code: fibres fire at a rate reflecting the frequency of the stimulus. Below 50 Hz, it appears that frequency is encoded solely by the frequency code. Frequency coding is also of particular importance when the sound is intense enough to saturate the neural firing rate (Section 4.2). Fibres of many characteristic frequencies will respond to an intense signal because it will be above threshold even for fibres with characteristic frequencies that are different from the signal frequency (although they will respond less vigorously). However, frequency information will still be encoded in the temporal firing pattern of all stimulated fibres.

4.2 Intensity coding

Information about stimulus intensity is encoded in two ways: the firing rates of neurons and the number of active neurons.

Firing-rate hypothesis

Intensity is assumed to be encoded by an increase in discharge rate of action potentials within the auditory system. As the stimulus gets more intense, the basilar membrane vibrates at a greater amplitude causing the membrane potential of activated hair cells to be more depolarized and this causes the nerve fibres that synapse onto the hair cells to fire at a greater rate. However for single fibres, the discharge rate increases only for a relatively small range of level changes. Figure 4.5 shows the results of recordings from an auditory nerve fibre in response to a stimulus of increasing intensity.

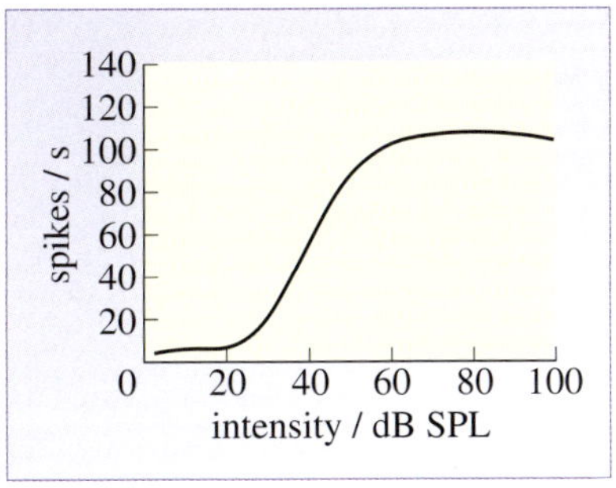

Figure 4.5 Intensity response function for a single auditory nerve fibre. The stimulus is a pure tone at the characteristic frequency of the neuron. As the intensity of the stimulus increases, so does the number of action potentials (spikes) generated per second.

❍ What does the figure show with respect to the responses of the fibre as a function of the sound level of the stimulus?

● The threshold of the fibre is about 25 dB SPL and the cell responds maximally to all sounds greater than about 65 dB SPL.

For this fibre a change in response with intensity only occurs over a range of about 40 dB, after which it no longer responds to increases in sound level with an increased firing rate: the fibre is said to be **saturated**. The range of sound levels between threshold and the level at which saturation occurs is called the **dynamic range**.

Humans are sensitive to a 140 dB dynamic range. Since a single fibre's discharge rate will only increase for a relatively small range of level changes (usually less than 35 dB), single fibre responses alone cannot encode changes in signal intensity. However, if intensity is determined by an increase in discharge rate of a large number of fibres with different response thresholds, then a large dynamic range could be accommodated. The most sensitive nerve fibres have response thresholds of about 0 dB SPL and characteristically have high rates of spontaneous activity. They produce saturating responses for stimulation at moderate intensities, about 40 dB SPL. At the opposite extreme, some afferent fibres display less spontaneous activity and much higher thresholds and give graded responses to intensities of stimulation in excess of 100 dB SPL. Activity patterns of most fibres are between these two extremes. Thus combining information from low-, medium- and high-threshold fibres may serve as the code for sound level.

Number of neurons hypothesis

In addition to an increase in firing rate of neurons with differing dynamic ranges, the inclusion of discharges from many fibres whose CFs are different from those of the stimulus may also help to account for the wide dynamic range of the ear. You know from Section 3.2.2 that in response to a pure tone stimulus the basilar membrane vibrates maximally at a given point. You should also be aware, however, that a pure tone will also cause vibration at points on the membrane adjacent to that of maximum stimulation. These vibrations are then reflected in the responses of the hair cells. Thus a pattern of excitation is produced in the auditory nerve such that fibres with a CF close to the input frequency will fire more strongly than those fibres whose CF is very different from the signal. Figure 4.6 shows a pattern of neural excitation along the basilar membrane that may be produced by a pure tone of 80 dB SPL (solid line). Assume that the neurons most excited by this stimulus are firing at a maximum rate so that any increase in intensity of the stimulus causes no increase in the firing rate of these cells. However, the cells with a CF either higher or lower than the stimulus frequency are not firing at their maximum level, so increasing the stimulus level causes an increase in firing rate of these cells. The effect of this is to broaden the excitation pattern as shown by the dashed curve. Thus, intensity could be encoded by how broad the excitation pattern may be to a given stimulus.

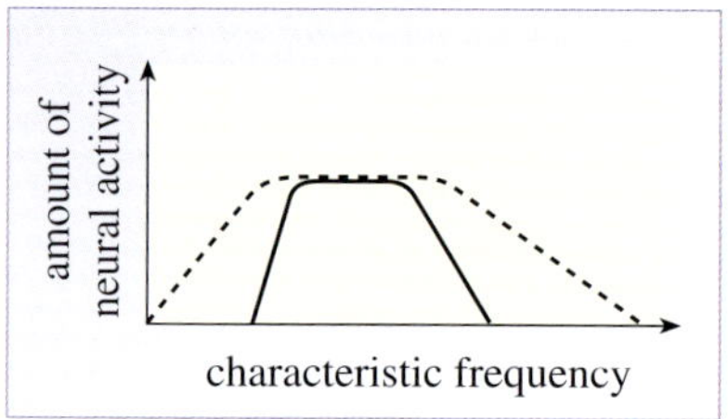

Figure 4.6 The proposed mechanism for the coding of intensity. An increase in stimulus intensity results in an increased firing rate of neurons with characteristic frequencies below or above the stimulus frequency, but no change in the firing rate of the neurons most sensitive to the stimulus, since these are already saturated.

READER

Now read Chapter 2 of the Reader, *The transformation of sound stimuli into electrical signals* by Robert Fettiplace.

Summary of Sections 4.0 to 4.2

Hair cells do not have axons and therefore do not generate action potentials.

The nerve that communicates with or innervates the hair cells along the basilar membrane is known as the vestibulocochlear nerve or VIIIth cranial nerve. The cochlear portion of the nerve contains afferent fibres that carry information in the form of action potentials from the organ of Corti to the brain, and efferent fibres that bring information from the cerebral cortex to the periphery.

Most of the afferent fibres connect to inner hair cells with which they make a many-to-one connection. The nerve cells that innervate the hair cells at the apex of the cochlea are in the middle of the nerve bundle while fibres from the base of the cochlea make up the outside fibres of the nerve bundle. The frequency-to-place conversion seen in the cochlea is therefore preserved in the auditory nerve.

There are two main theories concerning the way in which the auditory system encodes the frequency of the signal: the frequency code and the place code. Evidence suggests that the frequency code operates for frequencies below 50 Hz whereas the place code operates at frequencies above 1000 Hz. Both appear to play a role for frequencies between these.

There are also two main theories regarding how the auditory system encodes intensity information: the firing rate of neurons and the number of neurons that fire.

4.3 The central auditory nervous system

Up till now we have dealt with the anatomy of the auditory periphery and how the basic attributes of sound are coded within the auditory periphery. A great deal of additional processing takes place in the neural centres that lie in the auditory brainstem and cerebral cortex. Because localization and other binaural perceptions depend on the interaction of information arriving at the two ears, we need to study the central auditory centres, since auditory nerves from the two cochleae interact only at the brainstem and cerebral cortex. This section deals with the structure and function of the central auditory nervous system (CANS).

4.3.1 The ascending auditory pathway

Within the brainstem almost all fibres of the auditory nerve synapse on cells of the **cochlear nucleus**. The relationship between the cochlear nucleus and higher auditory centres is shown in Figure 4.7 (overleaf). This figure is highly schematic and simplified and shows only the main tracts and nuclei of the CANS, although other nuclei exist.

Once they leave the cochlear nucleus, most of the axons of the cochlear nucleus cells cross over to the opposite side (contralateral side) of the brain (Figure 4.7). This means that most of the auditory information processed by each half of the brain comes from the ear on the other side of the head. You will see in the next block that this is in contrast to that found in the visual system, where ganglion cell fibres either cross or stay on the same side of the brain in equal proportions. Both crossed and uncrossed fibres from the cochlear nuclei synapse in the area of the brainstem called the **superior olivary complex**. This is the first place in the ascending pathway to receive information from both ears. Neural impulses are transmitted from the superior olivary complex to the **inferior colliculus** through and/or around the **lateral lemniscus** (some fibres synapse in the lateral lemniscus but most travel through it to the inferior colliculus), from there to the **medial geniculate body** and finally to the **auditory cortex**. The location of the auditory cortex on the surface of the brain is shown in Figure 4.8 (overleaf).

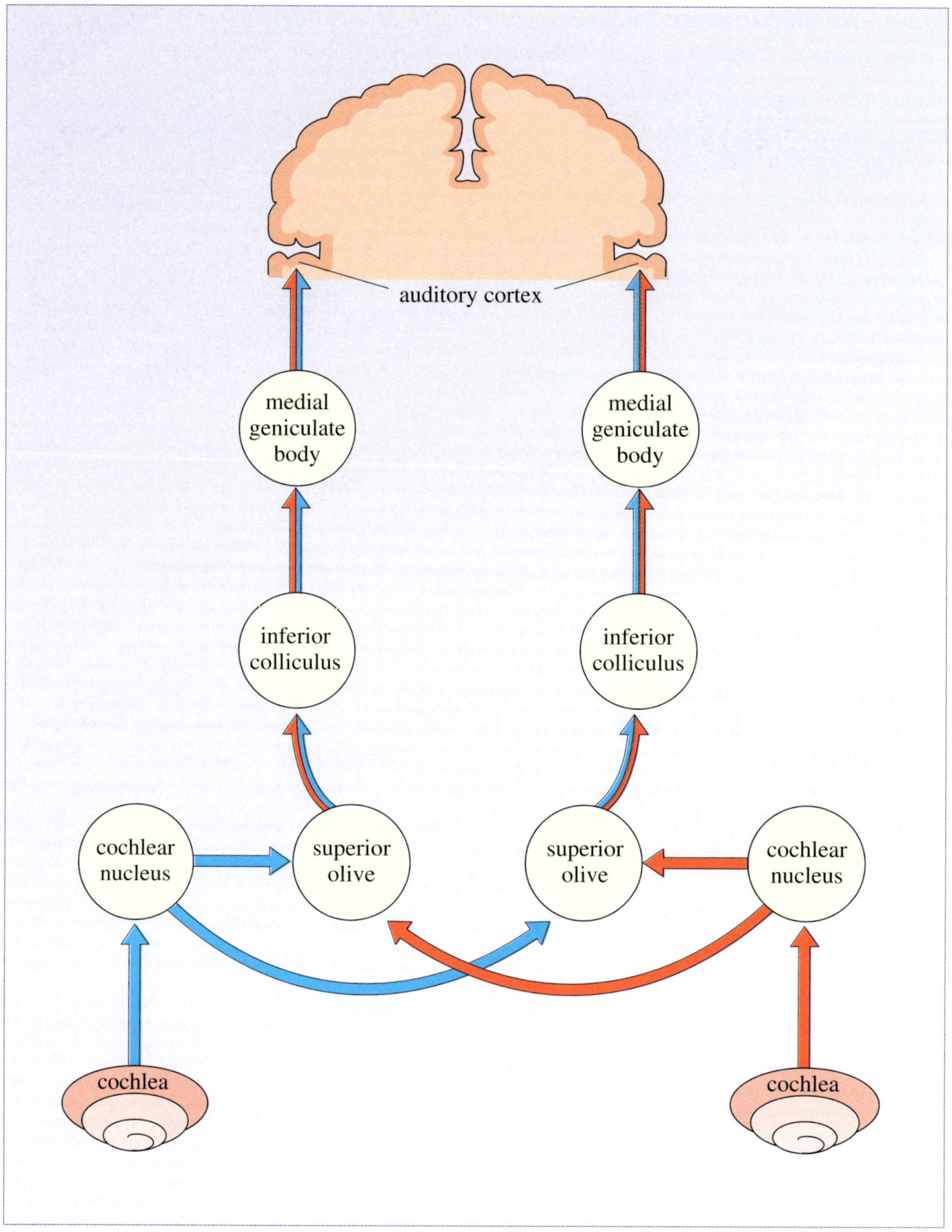

Figure 4.7 Highly schematic diagram of the bilateral central auditory pathway. The main pathways and nuclei are shown for both cochleae. Binaural stimulation occurs at the superior olive and all regions above.

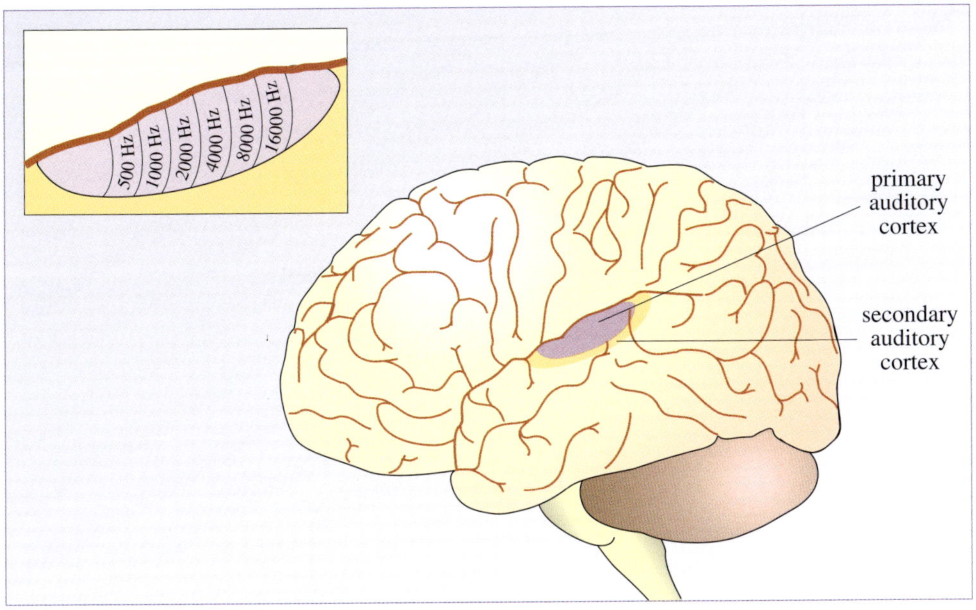

Figure 4.8 The primary auditory cortex in humans.

4.3.2 Coding of information in the higher auditory centres

We have seen that in the cochlear nerve, information about sound intensity is coded for in two ways: the firing rates of neurons and the number of neurons active. These two mechanisms of coding signal intensity are found throughout the auditory pathway and are believed to be the neural correlates of perceived loudness. The tonotopic organization of the auditory nerve is also preserved throughout the auditory pathway; there are tonotopic maps within each of the auditory nerve relay nuclei, the medial geniculate nucleus (MGN) and the auditory cortex. The MGN is shown in Figure 2.26 (where it is labelled the medial geniculate body) and Figure 2.27 (sub-divisions of the thalamus) of Block 2. Conversion from frequency to position that originates on the basilar membrane is maintained all the way up to the auditory cortex. One source of information about sound frequency is therefore derived from tonotopic maps; the location of active neurons in the auditory nuclei and in the cortex is an indication of the frequency of a sound. Phase locking as a means of frequency coding is also present in centres further along the pathway.

There are, in fact, two distinct pathways that occur in the CANS:

1. The '*what*' pathway which is monaural and receives information from only one ear. This pathway is concerned with the spectral (frequency) and temporal (time) features of a sound and is hardly concerned with the spatial aspects. It focuses mainly on identifying and classifying different types of sound.
2. The '*where*' pathway which is binaural and receives information from both ears. It is involved in the localization of a sound stimulus.

Despite the apparent dichotomy of these two processing pathways, the same types of acoustic cues may be important for the analysis that occurs in each. For example, spectral information is used in the 'where' pathway for determining a sound's elevation; and temporal information, used for our perception of frequency in the 'what' pathway, is also used in the 'where' pathway for determining a sound's horizontal location.

The 'what' pathway

The main nucleus involved in the 'what' pathway is the cochlear nucleus which has three main components, each of which is tonotopically organized; cells with progressively higher characteristic frequencies are arrayed in an orderly progression along one axis (Figure 4.9). The cochlear nuclei contain neurons of several types, each of which encodes a specific parameter of a stimulus (frequency, intensity, time): **stellate cells** encode *stimulus frequency* and *intensity*, **bushy cells** provide information about the *timing* of acoustical stimuli, and are involved in locating sound sources along the horizontal axis, and **fusiform cells** are thought to participate in the localization of sound sources along a vertical axis.

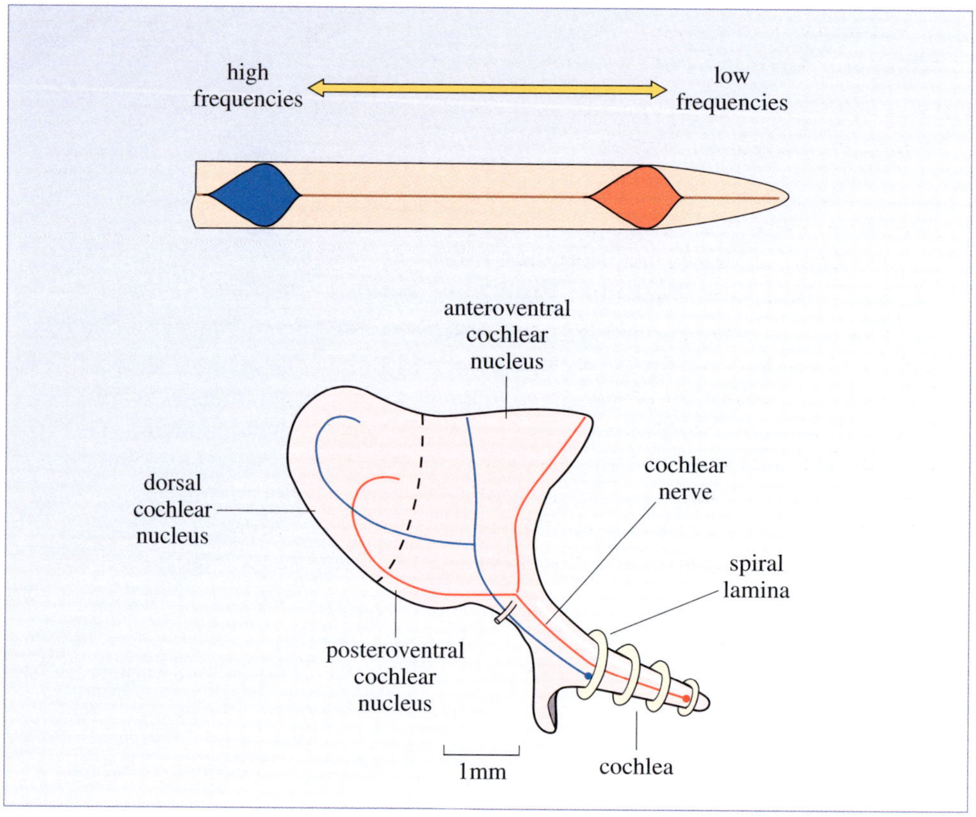

Figure 4.9 The representation of stimulus frequency in the cochlear nucleus. Stimulation with two sounds of different frequency causes vibration of the basilar membrane at two different positions (top). This in turn excites two distinct populations of afferent fibres, which project onto the cochlear nucleus in an orderly fashion.

The 'where' pathway

The 'where' pathway involves the ventral cochlear nuclei, the superior olivary complex and the inferior colliculus. The superior olivary complex is composed of the lateral superior olive (LSO) and the medial superior olive (MSO).

The neurons in the superior olivary complex are the first brainstem neurons to receive strong inputs from both cochleae and are involved in sound localization.

The MSO receives excitatory inputs from the cochlear nuclei on both sides and is tonotopically organized. It is involved in the localization of sound in the horizontal plane by processing information about auditory delays. Units in the MSO increase their firing rate in response to sounds from both ears as opposed to one ear, and these excitatory–excitatory (EE) units will increase their discharge rate further in response to sounds that reach both ears with a certain delay. In other words, a unit will discharge at the greatest rate when there is a particular interaural delay. This aids in localizing sound in the horizontal plane.

The LSO is also involved with sound localization but instead of using interaural time delays, it employs intensity differences to calculate where a sound originated. Information from the ipsilateral (same side) inputs to the LSO is usually excitatory and results in an increase in discharge rate of the neuron. Contralateral stimulation of the LSO is usually inhibitory. Thus, stimulation from both ears may decrease the firing rate of the neuron relative to the firing rate when only the ipsilateral ear receives sound. These excitatory–inhibitory (EI) units discharge with a few spikes when there is approximately equal stimulation of both ears and discharge rate increases as a function of changing the interaural level difference. The LSO therefore appears to form a network for processing interaural level differences, which are used to determine the location of sound sources.

The use of interaural time and intensity differences in sound localization will be dealt with in more detail in Section 5.4.

The inferior colliculus is part of the tectum and is the most prominent nucleus in the brainstem. It receives inputs from the olivary complex and the cochlear nucleus. Units in the inferior colliculus appear to be mainly EI units although there are EE units as well. They are tonotopically organized in sheets of cells (as in the cochlear nucleus). Cells in different parts of the inferior colliculus are either monaural, in that they respond to input from one ear only, or binaural, responding to bilateral stimulation. Both the spectral processing that takes place in the cochlear nucleus and the binaural processing that occurs in the olivary complex are seen in the inferior colliculus. In fact, the inferior colliculus is the termination of nearly all projections from brainstem auditory nuclei. It is therefore a 'watershed' for information processing where the 'what' and 'where' pathways converge on a single tonotopic map. Outputs of the inferior colliculus project mainly to the medial geniculate nucleus.

The medial geniculate nucleus is also tonotopically organized. Neurons with the same characteristic frequency are arrayed in one layer, so that the nucleus consists of a stack of neural laminae that represent successive stimulus frequencies. Sensitivity to interaural time or intensity differences is maintained. Axons leaving the MGN project to the auditory cortex. The neural responses of cortical cells in response to sound have been studied extensively in primates. In general, neurons are relatively sharply tuned for sound frequency and possess characteristic frequencies covering the audible spectrum of frequencies. In electrode penetrations made perpendicular to the cortical surface, the cells encountered tend to have similar characteristic frequencies, suggesting columnar organization on the basis of frequency, the so-called ice-cube model of the auditory cortex (Figure 4.10 overleaf). Although most of the neurons in the primary auditory cortex are sensitive to stimulation through either ear, their sensitivities are not identical. Instead the cortex is divided into alternating strips of two types. Half of these strips contain EE neurons and respond more to stimulation from both ears than to either ear separately, and the other half consist of EI neurons which are stimulated by unilateral input but inhibited by stimulation from the opposite ear. The strips of EE and EI cells run at right angles to the axis of tonotopic mapping so that the primary auditory cortex is partitioned into columns responsive to every audible frequency and to each type of interaural interaction.

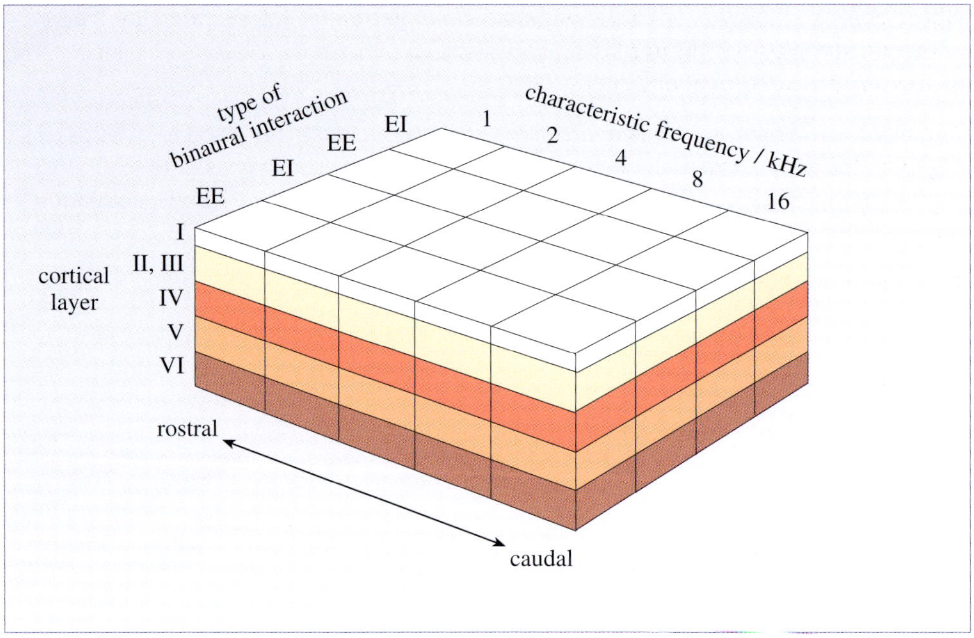

Figure 4.10 Ice-cube model of the auditory cortex.

4.3.3 The descending auditory pathway

The auditory system transmits information from the cochlea to the auditory cortex. Another system follows a similar path, but in reverse, from the cortex to the cochlear nuclei. This is the descending auditory pathway. In general, the descending pathway may be regarded as exercising an inhibitory function by means of a sort of negative feedback. It may also determine which ascending impulses are to be blocked and which are allowed to pass to other centres in the brain. The olivocochlear bundle, which arises from the olivary complex, is involved in sharpening or otherwise modifying the analysis that is made in the cochlea.

READER

Now read Chapter 3 of the Reader, *From cochlea to cortex* by Carole Hackney and Chapter 4, *Imaging central auditory function* by Alan Palmer and Debbie Hall.

These two chapters cover the most complex part of the auditory system, so you should not panic if you find them hard going; this is only to be expected. The best approach is probably to skim through each chapter first before reading them again more carefully.

Summary of Section 4.3

Fibres of the cochlear nerve synapse on the cells of the cochlear nuclear complex which is the first station of the central auditory pathway. From here signals are sent to the superior olivary complex, the inferior colliculus, lateral lemniscus, medial geniculate nucleus and finally the auditory cortex. The central role of the auditory cortex is the processing of complex sounds.

Each cochlear nuclear complex receives input from only one ear. In the cochlear nuclear complex are several different neural types that are responsible for extracting information about the spectral and temporal features of incoming sound.

Neurons in the superior olivary complex (SOC) are the first to receive input from both ears and are thought to play an important role in sound localization. The SOC processes information about interaural delays and intensities.

The inferior colliculus (IC) is a site for convergence of information. IC cells are organized in layers called sheets and within each sheet there appears to be a segregation of the EE and EI inputs. More complex aspects of a sound signal are processed in the IC and further features are extracted.

A tonotopic representation of frequency is seen at all levels of the auditory pathway.

Question 4.1

Describe how phase locking transmits information about the frequency of a sound (include the volley principle).

Question 4.2

The place theory of frequency coding is regarded as being of minimal use for very low frequencies. Why is this?

Question 4.3

Describe how the movement of the basilar membrane provides the brain with information about a signal's (a) frequency and (b) intensity.

Question 4.4

Use diagrams similar to Figure 4.4 to illustrate how the firing pattern of auditory neurons connected to two different (but close together) positions on the basilar membrane could encode information about the frequency and intensity of the following signals: (a) a low-intensity low-frequency tone; (b) a low-intensity high-frequency (but < 1 kHz) tone; (c) a high-intensity low-frequency tone; and (d) a high-intensity, high-frequency (but < 1 kHz) tone.

Question 4.5

Describe the method used to determine the characteristic frequency of a single auditory nerve fibre?

Question 4.6

Where in the higher auditory centres does binaural processing of information begin? What is the nature of the information used that enables us to localize sounds? How does the operation of excitatory and inhibitory inputs enable the auditory system to use this information?

Activity 4.1 *The Senses* CD-ROM and Hearing animation

This is a good time to take a break from reading and look at the hearing section of *The Senses* CD-ROM. You should also undertake the CD-ROM activities associated with this section, details of which can be found in the Block 3 *Study File*.

STUDY FILE

5 Auditory perception

We have learned so far that physical energy from the environment is transduced into electrochemical messages that affect the nervous system and give rise to psychological experiences, that is, produce sensations and perceptions. Sensation refers to the initial process of detecting and encoding environmental energy. The first step in sensing the world is performed by receptor cells, which in the case of hearing are the hair cells in the cochlea. Perception on the other hand, generally refers to the result of psychological processes in which meaning, relationships, context, judgement, past experience and memory all play a role. In many meaningful environmental encounters however, it is difficult to make such a clear distinction between sensation and perception. For example, when we hear a tune are we aware of any isolated tonal qualities of the notes, such as pitch and loudness, distinct from the melody? In most instances, perception and sensation are unified, inseparable psychological processes. In the next section we will look at an essential tool that has been used to study the quantitative relationship between environmental stimulation (the physical dimension) and sensory experience (the psychological dimension). We will then go on to examine the sensory or psychological effects produced by simple sounds and finally, look briefly at the reception of sound as meaningful information that allows us to perceive spatial features such as localization.

5.1 Psychophysics

Psychophysics is the oldest field of the science of psychology. It stems from attempts in the nineteenth century to measure and quantify sensation. It attempts to quantify the relationship between a stimulus and the sensation it evokes, usually for the purpose of understanding the process of perception. Historically, psychophysics has centred around three general approaches. The first involves measuring the smallest value of some stimulus that a listener can detect – a measure of sensitivity known as a **threshold**. The second is discrimination, where the subject is presented with two or more stimuli (e.g. two tones of different frequency) and then asked whether the stimuli are different. The third approach involves directly asking the listener about the stimulus. These are usually called scaling procedures.

The stimuli used in psychophysical tests can be varied along a number of dimensions. For example if the stimulus is light, it could vary in wavelength, size or shape; if it is sound it could vary in intensity, frequency or duration, etc. The response to the stimulus may be a verbal report ('yes I see it', 'no I don't see it', 'these two appear the same') or a mechanical response, such as pressing a button.

To estimate absolute thresholds and discrimination abilities, two basic classical psychophysical methods are used. The first of these is the **method of limits** and the second, the **method of constant stimuli**.

5.1.1 Absolute thresholds

The **absolute threshold** or **absolute limen** is the smallest value of a stimulus that an observer can detect. The concept of an absolute threshold assumes there is a precise point on the intensity or energy dimension that, when reached, becomes just perceptible to the observer and he or she responds 'yes – I can detect the stimulus'. It follows that when the stimulus is one unit weaker it will not be detected. If this were the case then some form of hypothetical curve, like the one shown in Figure 5.1, would be the result. However this rarely happens, as illustrated in the following example where an **auditory threshold** is derived using a traditional psychophysical method, the method of limits.

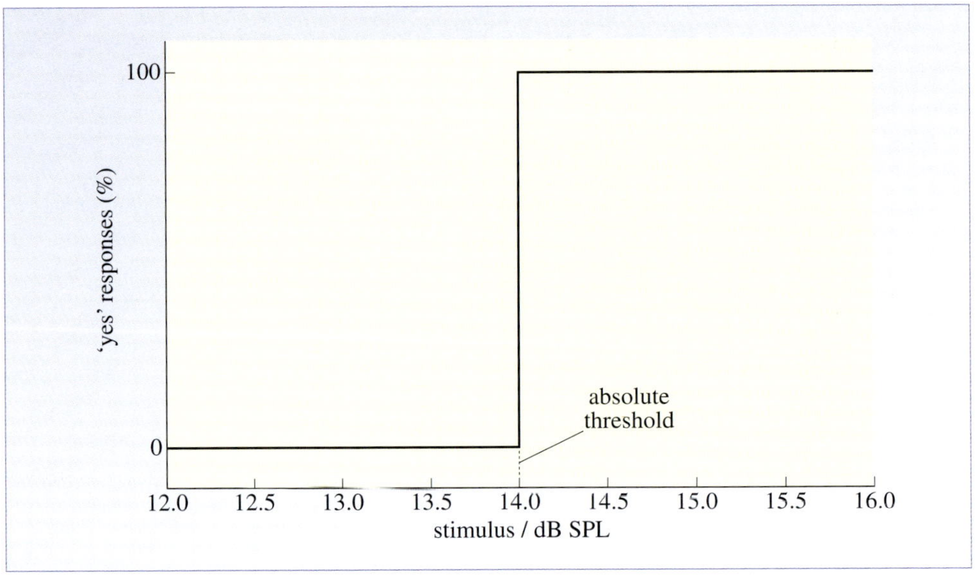

Figure 5.1 A hypothetical curve linking stimulus intensity to absolute threshold. The vertical axis plots the number of trials in which the subject responds 'yes', he/she can detect the stimulus. The threshold value is 14.0 dB SPL, i.e. below 14.0 dB SPL the stimulus is not detected.

Method of limits

To determine an auditory threshold using the method of limits, one would begin with an undetectable stimulus and then gradually increase the intensity until the subject detects it. Results from a hypothetical method of limits study are shown in Table 5.1. Stimulus intensity is shown in the first column and the subject's response to each stimulus is listed under 'trial 1'. Only when the stimulus was 14 dB SPL, did the subject respond 'yes' (I can hear the stimulus). The threshold for hearing the stimulus therefore lies somewhere between 13.5 and 14.0 dB SPL. Column 3 shows a repeat of the experiment. You can see that the responses of the subject in trial 2 were not the same as in trial 1. In this case the subject failed to detect the stimulus at a level of 14 dB SPL. As the experiment is repeated (trial 3 and trial 4) the responses differ from trial to trial. Loud stimuli, at intensities of 15 dB SPL and above are always heard whereas very soft stimuli (13 dB SPL and below) are never heard. Between these extremes, responses vary and are heard only a certain percentage of the time: stimuli at a level of 14.5 dB SPL were heard in 3 out of 4 trials (75%), whereas a stimulus level of 13.5 dB SPL was heard in only 1 out of 4 trials (25%). The reason why responses may differ is because actual thresholds change from trial to trial or because there is a variable amount of extraneous 'noise'. We shall return to this problem later. The percentage detection for each stimulus is shown in the last column. If we plot the percentage of stimuli detected against stimulus intensity we get a graph similar to that shown in Figure 5.2 – a smooth S-shaped curve known as a **psychometric function**. It is usual to define the threshold stimulus as that stimulus intensity corresponding to a 50% detection on the psychometric function.

❍ In this case, what is threshold for our subject?

● According to the graph, a 50% detection corresponds to a signal intensity of 14.0 dB SPL.

Table 5.1 Results from a hypothetical method of limits study.

Stimulus / dB SPL	trial 1	trial 2	trial 3	trial 4	% detection
12.0	N	N	N	N	0
12.5	N	N	N	N	0
13.0	N	N	N	N	0
13.5	N	N	Y	N	25
14.0	Y	N	Y	N	50
14.5	Y	Y	Y	N	75
15.0	Y	Y	Y	Y	100
15.5	Y	Y	Y	Y	100
16.0	Y	Y	Y	Y	100
16.5	Y	Y	Y	Y	100
17.0	Y	Y	Y	Y	100

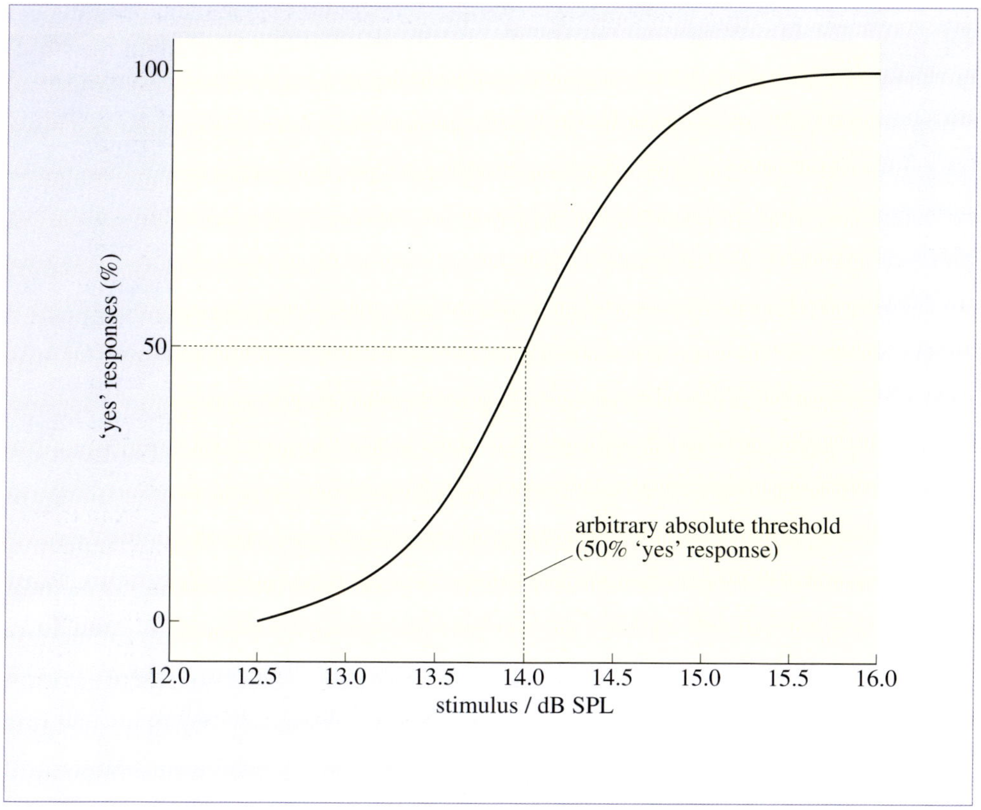

Figure 5.2 A typical threshold function. By convention, absolute threshold is defined as the intensity at which the stimulus is detected 50% of the time.

Although very useful, the method of limits is open to various sources of bias and error. One of its drawbacks is that the change in stimulus intensity (increase or decrease) is orderly and regular.

- ❍ How may this affect the observer?
- ● At any point the subject knows how intense a stimulus to expect next. As the series of presentations progresses, the expected intensity changes and the subject knows that the next stimulus will be more (or less) intense than the previous one. This could bias him or her to report a 'yes' when in fact the stimulus cannot be heard.

Method of constant stimuli

This method is similar to that described above but has two advantages over the method of limits. The first is that it's designed to overcome bias inherent in presenting stimuli in a set order. This is done by randomizing the order of presentation of stimuli. The subject therefore has no way of anticipating the intensity of the next stimulus (it could be softer or louder than the preceding one). In the table, the stimuli would be presented in a random order: for example, 13 dB SPL, 17 dB SPL, 12.5 dB SPL, etc. until each of the intensities is presented a sufficient number of times. Once again a psychometric function is generated and the intensity of the stimulus value detected on 50% of the trials is used as the measure of absolute threshold.

The method of constant stimuli has a second advantage over the method of limits in that the experimenter can get an estimate of the listener's bias by including 'catch' or 'blank' trials in the sequence. That is, occasionally the experimenter presents no stimulus to the listener. The listener does not know this has happened, and so the response is still 'yes' or 'no'. If there is a bias towards either response, then the proportion of 'yes' responses in the blank trials will reflect this bias.

5.1.2 Differential sensitivity

Absolute thresholds represent only one type of threshold; one could also ask whether the subject can detect a *difference* between two stimuli. The threshold for detection of difference is called a **difference threshold** or **difference limen** (**DL**). The difference threshold is a measure of the smallest detectable difference between two stimuli. Basically it answers the psychophysical question: 'How different must two stimuli (e.g. two weights, two colours, two sounds) be from each other in order to detect them as different stimuli?'

The difference threshold, like the absolute threshold described in the beginning of this section, is a derived statistical measure; it is the difference in magnitude between two stimuli, usually a standard (S) and a comparison stimulus (T), that is detected 50% of the time. For example, if two tones of the same intensity are presented to a listener, the listener will generally report that they are equal in loudness. However, as the intensity of one of the tones is gradually increased, an intensity difference between the tones will be reached at which they will be judged different in 50% of trials. The magnitude of this difference specifies the difference threshold; that is, the amount of change in a stimulus necessary to produce a **just noticeable difference (JND)** in sensation. If the magnitude of a stimulus, say a sound, is 100 dB SPL, and the sound has to be increased to 110 dB SPL in order to be perceived as different, then the JND equals 10 dB.

The measurement of a JND can be made using any of the classical methods discussed above. A psychometric function similar to that shown in Figure 5.2 is generated except that the horizontal axis is ΔI instead of I. ΔI is the increment in intensity that, when added to the stimulus intensity I, produces a JND or the smallest detectable increment. From the psychometric function, a threshold value, the difference threshold, can be deduced. This difference threshold is the ΔI appropriate to the standard (S) used. For a different S, a different psychometric function is generated, and a different ΔI derived. By measuring the ΔI for a large number of standards you can develop a function that describes how the JND changes for different levels of stimulation.

Weber's Law

Pioneering work on the relationship between ΔI and S was done by the German physiologist, Ernst Weber in the 1830s. Weber found that the increment in stimulation required for a JND was proportional to the size of the stimulus. Weber had subjects lift a small 'standard' weight (S) and then lift a slightly heavier 'comparison' (T) weight and judge which was heavier. He found that when the difference between the standard and comparison weights was small, the subjects found it difficult to detect a difference between the weights, but could easily detect large differences – not really suprising. However, he also found that the size of the JND depended on the size of the standard weight. For example, the JND for a 100 g standard weight was 5 g ($=\Delta I$). In other words, the subject could tell the difference between a 100 g standard and a 105 g comparison weight but couldn't tell the difference between a 100 g weight and a comparison weight less than 5 g heavier. In contrast, the JND for a 200 g weight was found to be 10 g ($=\Delta I$); the subject could only detect a difference between a standard 200 g weight and a comparison weight of 210 g or more. Thus, as the magnitude of the stimulus increases, so does the size of the JND.

Research on a number of senses, including hearing, has shown that the JND is larger for larger standard stimuli and that, over a fairly large range of intensities, the ratio of JND to the standard stimulus is constant, i.e.

$$\Delta I / S = \mathrm{K} \qquad (5.1)$$

This is called **Weber's law** which states that the bigger the stimulus, the bigger the increment needed for a change to be detectable. K is called the Weber fraction. Applying the equation to our example of lifted weights, we find that for a 100 g standard, $\mathrm{K} = 5 / 100 = 0.05$.

❍ What does K equal for a 200 g weight?

● $\mathrm{K} = 10\ \mathrm{g} / 200\ \mathrm{g} = 0.05$.

Therefore the Weber fraction is constant. What this means is that if I increases, ΔI must increase correspondingly. So, if the standard intensity is low, the increment of change necessary to produce a JND is correspondingly small; by contrast, if the initial intensity is high, the stimulus increment necessary for the JND is correspondingly large.

The Weber fraction holds for most senses, as long as the stimulus intensity is not too close to threshold.

5.2 The perception of loudness

The human ear has incredible absolute sensitivity and dynamic range. The most intense sound we can hear without immediate damage to the ear is at least 140 dB above the faintest sound we can just detect. This corresponds to an intensity ratio of 100 000 000 000 000 : 1. In this section, we examine how the loudness of a sound can be measured and how the perception of loudness is affected by the intensity and duration of the signal.

5.2.1 Absolute thresholds

You know from Section 5.1.1 that the absolute threshold is the smallest value of some stimulus that a listener can detect. In order to investigate our perceptual capabilities, it is useful to generate an absolute threshold curve, which relates the frequency of a signal to the intensity at which it can be detected by the ear. Figure 5.3 is a plot of the thresholds of hearing for a range of frequencies.

❍ From the graph, describe the relationship between signal frequency and the threshold of audibility of a signal.

● Auditory thresholds are lowest for tones around 3000 Hz and increase for tones of higher and lower frequency.

This means that people are therefore most sensitive to tones of frequencies around 3000 Hz, with sensitivities decreasing for tones that are either higher or lower in frequency. There will be very high and very low frequencies to which, no matter how intense the sinusoidal wave, the auditory system is insensitive. These frequency limits define the bounds of the auditory system's sensitivity to frequency.

In order to generate an audibility curve like that shown in Figure 5.3, you would determine the level required for a listener to detect the presence of a sinusoidal wave at each of many frequencies. One method of doing this involves delivering the sound using loudspeakers and measuring the sound pressure at the entrance to the auditory meatus at threshold. A threshold measured in this way is known as a minimum audible field (MAF). In contrast when sounds are delivered through earphones the threshold measured is called the minimum audible pressure (MAP). MAP thresholds are plotted as a function of frequency in Figure 5.3.

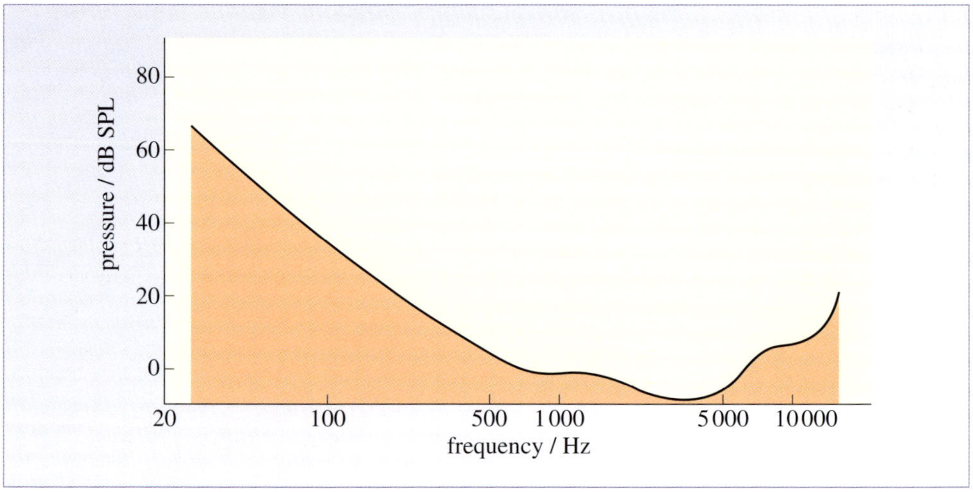

Figure 5.3 Human auditory thresholds as a function of frequency. Sounds that fall in the shaded region below the curve are below threshold and therefore inaudible.

The threshold sound levels displayed in Figure 5.3 produce extremely small physical displacements at both the tympanic and basilar membranes. In humans, the sound level of frequencies to which we are most sensitive cause movements of the basilar membrane of about 0.2 nm – about the diameter of two hydrogen atoms.

Figure 5.3 shows auditory thresholds for young people. As we grow older, we become less sensitive to stimuli of all frequencies, but the maximum hearing losses occur for high frequency tones.

❍ Can you think of a reason why this may be the case?

● Remember that high frequencies stimulate the basal end of the basilar membrane, even though the peak in displacement is towards the apex. This part of the membrane in fact is stimulated by all frequencies to some extent and the hair cells in this region have a greater potential for being 'worn out'.

5.2.2 The relationship between loudness and intensity

The loudness of an auditory stimulus is a psychological, not physical attribute of the stimulus. The physical attribute of sound that is most closely correlated with loudness is intensity. So loudness is the listener's subjective description of the intensity of the stimulus. As you know, we are not equally sensitive to sounds of all frequencies so perceived loudness of a tone in fact depends on frequency as well as intensity. Two sounds can have the same physical sound pressure levels but if they are of different frequencies, they are often perceived as having different loudness.

How do we measure loudness? To do so, we have to relate a subjective quality such as loudness to a physical quantity such as sound pressure level. One way of doing this is to generate a plot of **equal loudness contours**. A 1000 Hz tone is set to some specific intensity and then the sound levels of other tones of different frequency that are equal in loudness to the 1000 Hz tone, are determined. For example, a subject may be presented with a 1000 Hz standard tone at 60 dB SPL and then asked to manipulate the intensity of a 2000 Hz tone until it matches the loudness of the 1000 Hz tone. The same 1000 Hz tone would then be compared with a 3000 Hz tone and the intensity manipulated till it matched the 1000 Hz tone in loudness. In this way, the intensity of tones at a variety of frequencies could be obtained so that all tones matched the loudness of the 60 dB SPL, 1000 Hz tone. These intensities are then plotted as a function of frequency to generate equal loudness contours as shown in Figure 5.4 (overleaf). The term that is used to describe or measure the loudness of a signal is known as a **phon**. The loudness in phons is the level in dB SPL of an equally loud 1000 Hz tone. So all tones judged equal in loudness to a 40 dB SPL, 1000 Hz tone have a loudness of 40 phons. The tones presented at levels such that they are equal in loudness to a 70 dB SPL, 1000 Hz tone all have loudness levels of 70 phons, and so on. The equal loudness contours in Figure 5.4 (overleaf) form the phon scale. Look at the contour labelled 30 in Figure 5.4. Any sound whose frequency and intensity lie on the contour sounds just as loud as any other sound on the contour, although the frequency and intensity of the two sounds will differ. So for example, a 60 Hz tone at a 65 dB intensity level and a 330 Hz tone at a 40 dB intensity level both sound as loud as a 1000 Hz tone at a 30 dB intensity level and all have a loudness level of 30 phons.

❍ What do you notice about the contours as loudness level increases from 0 to 120 phons?

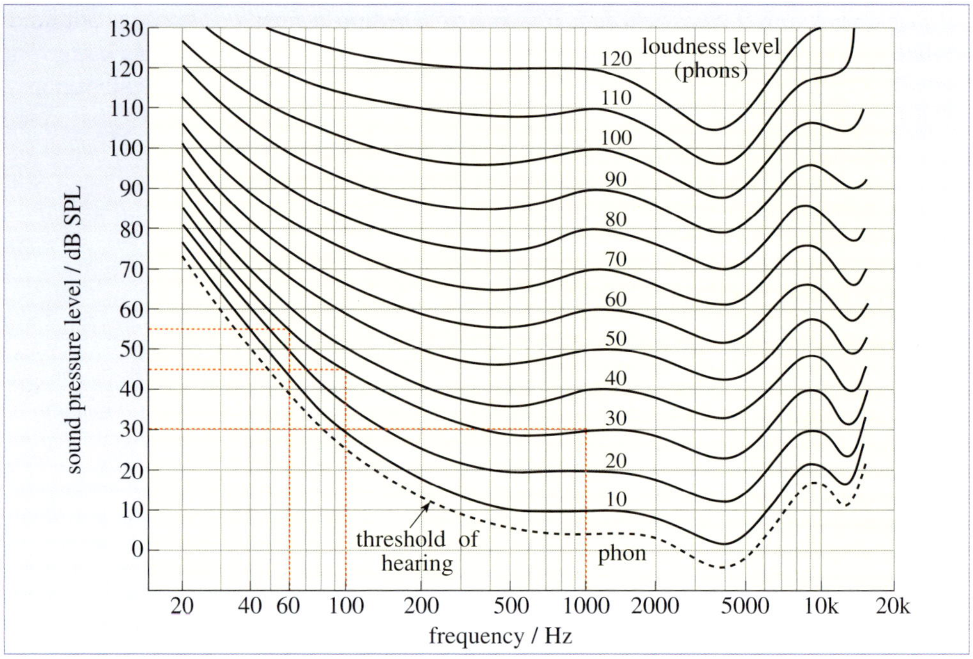

Figure 5.4 Equal loudness contours. The curves show the sound intensities (in dB SPL) required at different frequencies, to produce sounds that appear equally loud; in other words they have the same loudness value, measured in phons. The phon scale has the SPL scale at its heart, and uses the same standard frequency of 1000 Hz. Thus, whatever the intensity (SPL) of a 1000 Hz tone, by definition that is also its loudness value (phons). For example, a 1000 Hz tone of 60 dB SPL must have a loudness of 60 phons. Because we are not equally sensitive to sound at different frequencies, for a sound to remain at 60 phons it would generally require a higher than 60 dB intensity as the frequency shifted away from 1000 Hz.

NB: Although in theory a 1000 Hz tone that was just inaudible (0 phons) would lie on the 0 dB line, in practice the average person's hearing is not quite so good. Their cut-off point is just slightly more intense than this; hence the dotted line not dipping to 0 dB at 1000 Hz.

- The contours change in shape and become much flatter.

What this indicates is that at high intensity levels, the frequency of a sound becomes less important in the perception of loudness. Look at the 120 phon contour. In order to sound equally loud, a 60 Hz tone, a 100 Hz tone and a 1000 Hz tone need to differ by a maximum of 8 dB in intensity (128 dB SPL, 125 dB SPL and 120 dB SPL respectively). This contrasts with the 30 phon contour where, in order to sound equally loud, the three tones differ by as much as 25 dB. This in effect means that we are relatively more sensitive to low-frequency tones than to high-frequency tones at high loudness levels. Below about 70 phons, low-frequency tones require a higher intensity to achieve comparable loudness with higher-frequency tones. This is especially true for sounds with frequencies below about 1000 Hz.

Many sounds, such as music, contain a range of frequencies, but the relative loudness in each frequency range may seem to change, depending on the overall sound intensity. For example, if you play music from a loudspeaker with the volume turned very low, you may feel that the deeper bass notes are not loud enough, or even completely missing. Looking at Figure 5.4 you will see why. We are relatively insensitive to low frequencies; in other words, to reach the 0 phon level low frequencies have to be several decibels more intense than high frequencies. However, if we turn the volume up high we may begin to think there is too much bass and that the sound is 'boomy'. This is because the louder sound curves are more flat, so the low frequencies are no longer at such a disadvantage.

5.2.3 Intensity discrimination

The smallest detectable change in intensity has been measured using a variety of psychophysical methods and various stimuli. Although the difference threshold depends on several factors including duration, intensity and the kinds of stimuli on which the measurement is made, Weber's law holds for most stimuli. In other words, the smallest detectable change is a constant fraction of the intensity of the stimulus. Expressed in dB, the minimum change in intensity that produces a perceptual difference is about 0.5 to 1.0 dB. However, for pure tones Weber's law does not hold in that discrimination, as measured by the Weber fraction, improves at high levels. For a 1000 Hz tone, the difference threshold ranges from 1.5 dB at 20 dB SPL to 0.3 dB at 80 dB SPL.

5.3 The perception of frequency

Although the perception of sound involves the interaction of frequency and intensity, many aspects of frequency reception can be analysed separately.

For normal or typical hearing, the limits of hearing for frequency fall between 20 and 20 000 Hz. Below 20 Hz only a feeling of vibration is perceived; above 20 000 Hz, only a 'tickling' is experienced.

5.3.1 The relationship between frequency and pitch

As well as loudness, the other most obvious characteristic of a sound is its pitch. **Pitch** is a subjective dimension of hearing. It is the sound quality most closely related to the frequency of a pure tone. High-frequency tones are perceived as being of high pitch while low-frequency tones are said to be of low pitch. The relationship between pitch and frequency is however, not a simple linear one. In order to investigate how the two are related, pitch has been assigned the arbitrary unit the **mel**. The pitch of a 1000 Hz tone at 40 dB SPL has been given a fixed value of 1000 mel. In order to determine the number of mels that are associated with different frequency tones, a subject is presented with a 1000 Hz tone and told that the pitch is 1000 mels. The subject is then asked to manipulate the frequency of a tone until that tone has a pitch that is one half as high as the 1000 mel tone. This tone is then assigned a value of 500 mel. The subject is then asked to find a frequency that is half the pitch of the 500 mel tone which is then assigned a value of 250 mel. In this way a function relating frequency to mels can be generated (Figure 5.5). Figure 5.5 shows that pitch

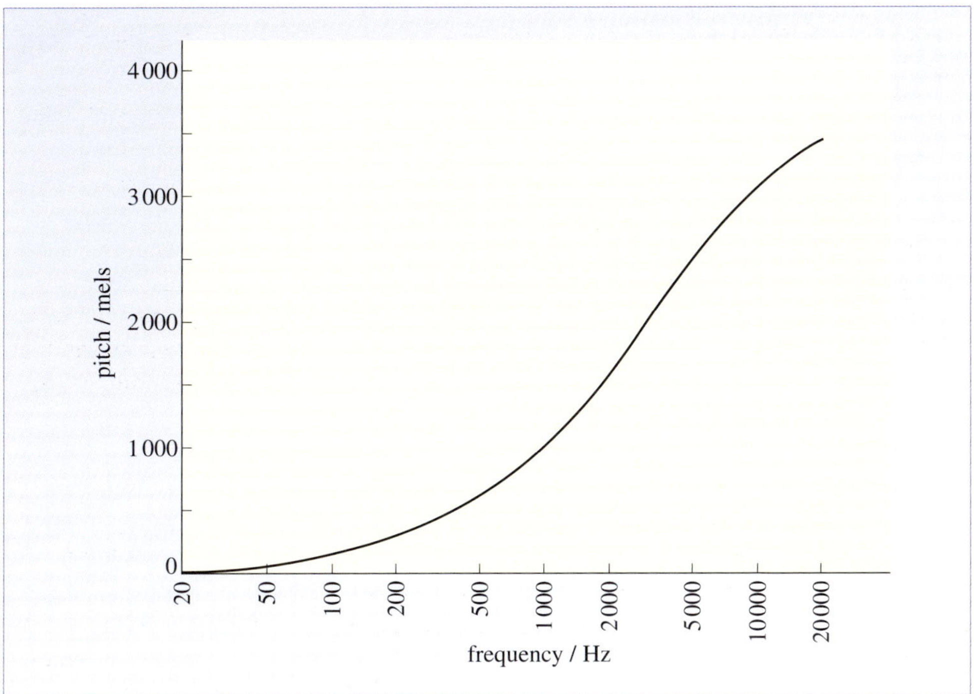

Figure 5.5 Pitch in mels plotted against frequency (in Hz). The curve shows that the perceived pitch of a tone varies with frequency.

is not related to frequency in either a linear fashion or a logarithmic fashion (note that frequency is plotted on a logarithmic scale); the relationship is more complex. In general, pitch increases more rapidly than frequency for tones below 1000 Hz and less rapidly for tones above 1000 Hz. That is, for frequencies above 1000 Hz a greater change in frequency is needed to produce a corresponding change in pitch.

5.3.2 Frequency discrimination

Some findings indicate that, for moderate loudness levels, humans can detect a frequency change of about 1 to 3 Hz for frequencies up to about 1000 Hz. Figure 5.6 shows a plot of the smallest frequency difference for which two tones can be discriminated for a number of reference tones. You can see from the figure that up to about 1000 Hz, the DL is between 1 and 3 Hz. In fact, for frequencies between 500 and 2000 Hz, discriminability is a constant fraction of the frequency to be discriminated. In other words, the Weber fraction ($\Delta F / F$) for this frequency interval remains constant, at approximately 0.002. Although this holds true for a wide range of intensities the intensity of the sound does affect the determination of the minimal discriminable change in frequency. The DL for frequency increases as the stimulus intensity decreases. In other words, as the intensity of the sound decreases, it is more difficult to detect it as being different from other sounds close to it in frequency.

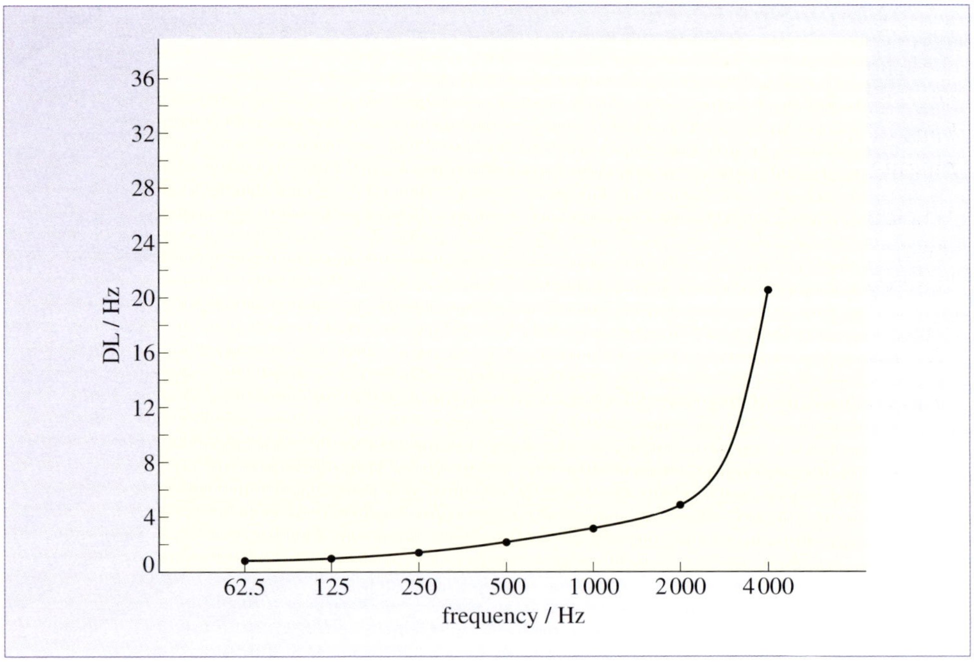

Figure 5.6 Difference limens for pitch as a function of frequency at a moderate loudness level.

5.3.3 Frequency selectivity

In preceding sections we examined two ways in which the auditory system may code frequency information: the place theory and phase locking. In this section we will look at the psychophysical evidence for place coding on the basilar membrane by examining the ability of the auditory system to resolve the components of sinusoidal waves in a complex sound – a phenomenon known as frequency selectivity.

The perception of a sound depends not only on its own frequency and intensity but also on other sounds present at the same time. You will all be familiar with the experience of one sound 'drowning out' another sound. For example, typical classroom sounds, created by movement, coughing, rustling of papers, make the instructor's voice difficult to hear. This phenomenon is called **masking**. Technically speaking, masking is defined as the rise in threshold of one tone (test tone) due to the presence of another (masker) tone.

It has been known for many years, that a signal is most easily masked by a sound having frequency components close to those of the signal. This led to the idea that our ability to separate the components of a complex sound depends on the frequency-resolving power of the basilar membrane. It also led to the idea that masking reflects the limits of frequency selectivity and provides a way to quantify it.

A masking experiment

The procedure for a masking experiment is shown in Figure 5.7. First, the threshold for hearing is determined across a range of frequencies (Figure 5.7a). Then, a masking stimulus is presented at a particular place along the frequency scale and while the masking stimulus is sounding, the thresholds for all frequencies are re-determined (Figure 5.7b). The masking stimulus used can be a pure tone or, more commonly **white noise**. A white noise stimulus is simply one that contains a band of frequencies with equal sound pressure at each frequency. It sounds something like the 'shhhhhhh' sound you can make by blowing air across your teeth when they are slightly separated. The band of frequencies used can vary in width. For example, a band of frequencies 90 Hz wide centred on 410 Hz would contain frequencies ranging from 365 to 455 Hz.

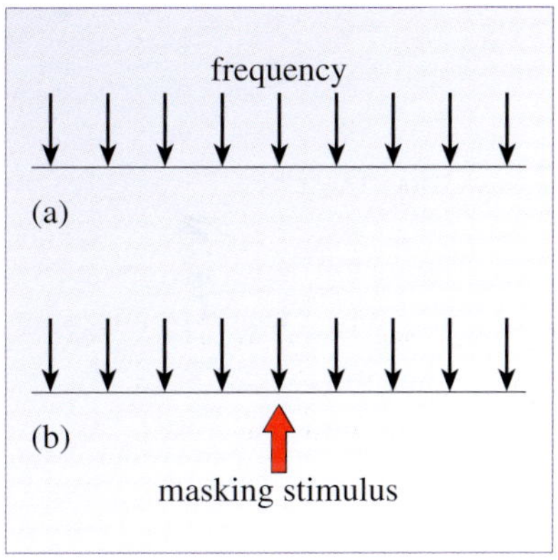

Figure 5.7 The procedure for a masking experiment. (a) The threshold is determined across a range of frequencies. Each arrow indicates a frequency where the threshold is measured. (b) The threshold is re-determined at each frequency (small arrows) in the presence of a masking stimulus (large arrow).

When thresholds are measured in the presence of a masking noise the original thresholds are raised. Figure 5.8 shows the result of an experiment using a 90 Hz band of noise centred at 410 Hz.

There are two things to note about Figure 5.8. First, the threshold increases most for frequencies near the frequencies in the masking tone. Second, the curve is not symmetrical; the masking effect spreads more to high frequencies than to low frequencies. So, lower frequencies mask higher-frequency sounds much more effectively than the reverse.

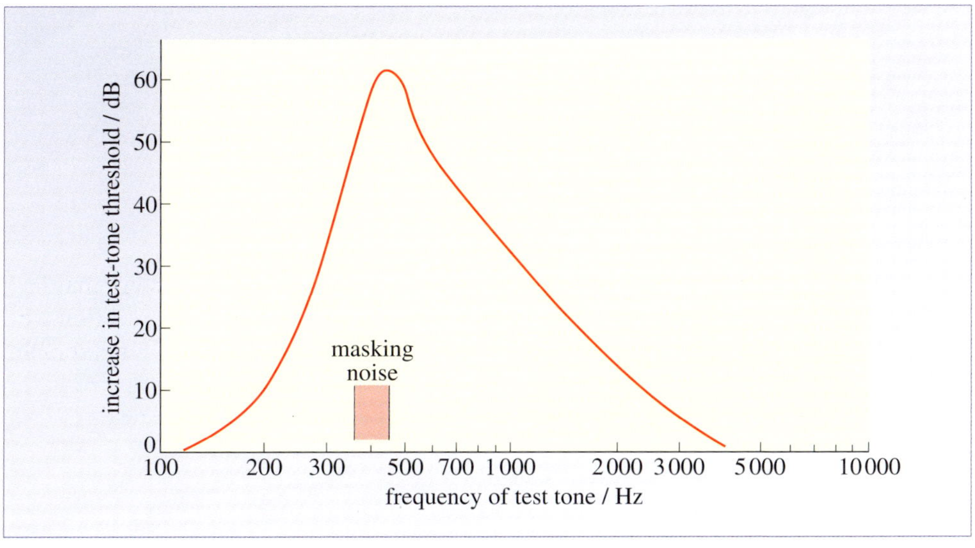

Figure 5.8 Results of a masking experiment. The red line indicates the amount that the threshold is raised in the presence of a masking noise centred at 410 Hz. So for a 410 Hz tone, the threshold is raised by about 60 dB above absolute threshold.

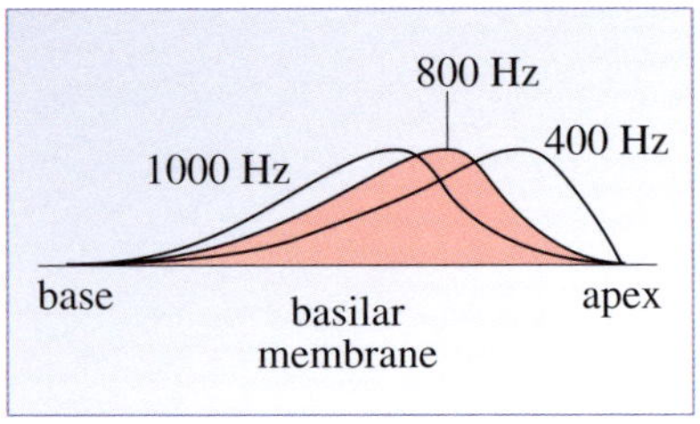

Figure 5.9 Vibration patterns on the basilar membrane caused by 400, 800 and 1000 Hz tones.

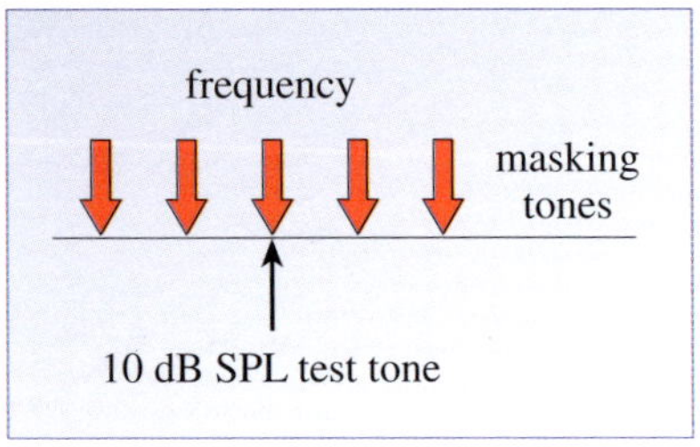

Figure 5.10 The procedure for measuring a psychophysical tuning curve. A 10 dB test tone (black arrow) is presented and then a series of masking tones (red arrows) are presented at the same time as the test tone. The psychophysical tuning curve is generated by determining the SPL threshold of the masking tones needed to reduce the perception of the test tone to threshold.

How is this masking effect explained?

Many masking effects can be explained in a very simplified way by analysing the interaction of displacement patterns in response to sound on the basilar membrane. Figure 5.9 shows the vibration patterns on the basilar membrane caused by a 400 Hz, 800 Hz and 1000 Hz tone. You can see that the vibration pattern of the 800 Hz tone overlaps those of the 400 and 1000 Hz tones. Note also that the pattern for the 800 Hz tone (which is shaded) almost totally overlaps the pattern for the higher frequency, 1000 Hz tone, but does not overlap the place of peak vibration of the lower frequency, 400 Hz tone. We would therefore expect the 800 Hz tone to mask the 1000 Hz tone more effectively than the 400 Hz tone. This is what happens, providing support for the place mechanism of frequency tuning on the basilar membrane.

Masking has also been used to determine psychophysical tuning curves. A low intensity tone, called the test tone, is presented throughout the experiment. A series of masking tones (also pure tones) are then presented, one at a time, with the test tone. One of the masking tones is the same frequency as the test tone and the others are higher or lower in frequency. The level of each masking tone is reduced until the test tone is just audible. To understand the rationale behind this experiment, imagine that the horizontal line in Figure 5.10 is the basilar membrane. Each masking tone, because it is a pure tone (single frequency), will cause vibration mainly at one point on the basilar membrane. If the masking tone is the same, or of similar frequency to the test tone, it will mask the test tone more effectively than a masking tone with a frequency far away from that of the test tone. The results of this kind of masking experiment are shown in Figure 5.11a. They show what we would expect according to the place theory of frequency coding on the basilar membrane: when the masking tone is the same frequency as the test tone it doesn't need to be very intense in order to mask the test tone. However, when the masking tone is higher or lower in frequency than the test tone, higher intensities are required to mask the test tone. You can compare the psychophysical tuning curves in Figure 5.11a with those obtained from auditory nerve fibres of a cat seen in Figure 5.11b.

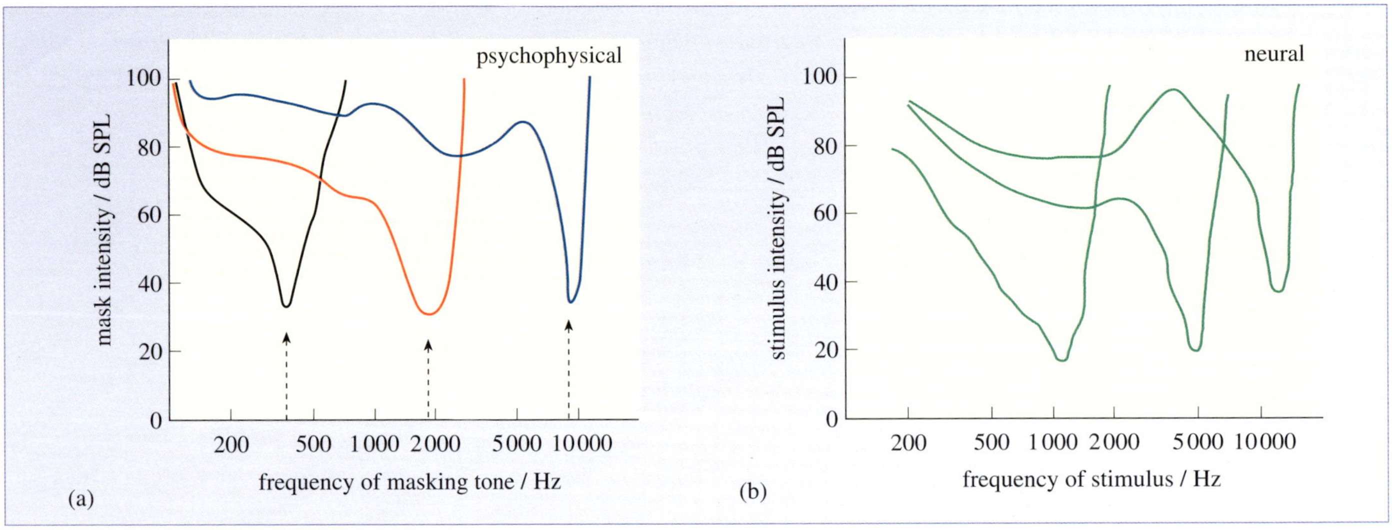

Figure 5.11 (a) Three human psychophysical tuning curves generated using the method described in Figure 5.10. The arrows show the frequency of three different test tones. You can see from the figure that when the masking tone is the same as, or close to, the test tone in frequency, the intensity of the masker needed to mask the test tone is low. (b) Three neural tuning curves showing the stimulus intensity needed to generate a constant response (firing rate) in the nerve fibre of a cat. Each curve represents a different auditory nerve fibre.

The close match between them suggests that both reflect the same process – a place code for frequency on the basilar membrane.

5.3.4 Signal duration

Since hearing is largely a matter of stimulus reception over time, we would expect time to influence the perception of sound. It has been known for many years that both absolute thresholds and the loudness of sounds depend upon signal duration. The studies of absolute threshold described earlier were all carried out with tone bursts of relatively long duration. For durations exceeding 500 ms, the sound intensity at threshold is roughly independent of duration. However for durations of less than 200 ms the sound intensity needed for detection increases as signal duration decreases. This also means that for sounds of less than 200 ms duration the intensity must be increased to maintain a constant level of loudness. Intensity also affects the perception of frequency. For example, if a tone of an audible frequency and intensity is presented for only a few milliseconds, it will lose its tonal character and will either be inaudible or be heard as a click. The length of time a given frequency must last in order to produce the perception of a stable and recognizable pitch is about 250 ms. We are also able to discriminate between tones of different frequencies when their duration is lengthened.

Summary of Sections 5.0 to 5.3

In this section we have described some of the quantitative relationships between the physical dimensions of simple sounds and their subjective psychological dimensions. The physical dimension of intensity, or pressure amplitude, given in decibels (dB), directly affects loudness. Frequency of pressure changes, in hertz (Hz), mainly determines pitch.

The lowest threshold value and hence the maximal sensitivity for humans is in the region of 3000 Hz.

The quantitative relationship between intensity and loudness is that loudness grows more slowly than intensity. Equal loudness contours indicate that humans are more sensitive to frequencies between 1000 and 4000 Hz than other frequencies within the hearing range. When intensity is held constant, sounds in the region of 3000 Hz appear louder than sounds of other frequencies. The minimum change in intensity of a sound that produces a perceptual difference is 1 to 2 dB.

A number of different mechanisms play a role in intensity discrimination. Intensity changes can be signalled by both changes in the firing rates of neurons at the centre of the excitation pattern, and by the spreading of the excitation pattern. In addition, cues related to phase locking may also play a role in intensity discrimination. This may be particularly important for complex stimuli, for which the relative levels of different components may be signalled by the degree of phase locking to components.

The relationship between frequency and pitch is investigated using the mel scale. This shows that pitch is not linearly related to frequency. Pitch increases more rapidly than frequency for tones below 1000 Hz and less rapidly for tones above 1000 Hz.

The difference threshold for frequencies up to 1000 Hz is about 3 Hz, whereas for frequencies between 1000 and 4000 Hz the Weber fraction remains constant at about 0.002. Intensity affects the difference threshold: the lower the intensity, the higher the difference threshold.

Masking experiments support the place theory of frequency selectivity on the basilar membrane.

Although the sound heard depends primarily on its frequency and intensity, both its pitch and its loudness are secondarily affected by the duration of the sound. Within limits, loudness and pitch recognition increases as the duration of a brief burst of sound is lengthened.

STUDY FILE

Activity 5.1 Perception: pitch and loudness, frequency difference limen and beats

It's time to take a break again and return to the CD-ROM. There are a number of activities which demonstrate some of the perceptual phenomena you have studied. Details of these activities are given in the Block 3 *Study File*.

5.4 Sound localization

While information about frequency and intensity is essential for interpreting sounds in our environment, sound localization can be of critical importance for survival. For example, if you carelessly cross the street, your localization of a car's horn may be all that saves you. Our current understanding of the mechanisms underlying sound localization suggests that we use different techniques for locating sources in the horizontal plane and vertical plane.

5.4.1 Localization of sound in the horizontal plane

❍ Imagine a sound source that is directly in front of you. All else being equal, will the sound reach each ear at the same time?

● Yes, if it is directly in front of you, since the distance it must travel to each ear is the same.

❍ Will the sound be equal in loudness at your two ears given that as sound travels over distance, it decreases in intensity?

● Yes, because it travels the same distance and therefore attenuates to the same extent before reaching each ear.

❍ Now imagine a sound source that is directly to one side of your head – say the left side. Which ear will receive the sound first?

● The left ear, because the distance the sound must travel to the left ear is shorter than to the right (it must travel over your head).

❍ At which ear will the sound be louder?

● At the left ear since it travels less distance and therefore attenuates less compared to the sound arriving at your right ear.

These two kinds of information, differences in intensity of sound at the two ears (**interaural intensity differences**) and differences in the time of arrival of sound at the two ears (**interaural time delay**) enable our auditory system to localize a sound source in the horizontal plane.

Interaural time delays: non-continuous sounds

The average distance between human ears is about 20 cm. Therefore, if a sudden noise comes at you from the right, perpendicular to your head, it will reach your right ear 0.6 ms before it reaches your left ear. For a sound coming from directly in front of you there will be no delay, and at angles between, the delay will be between 0 and 0.6 ms. Therefore there is a simple relationship between the location of the sound source and the interaural delay. It is this delay that enables us to localize the source of a sound in the horizontal plane. How does the auditory system encode information about interaural time delays? It is thought that the mechanism involves a series of **delay lines** and **coincidence detectors,** as illustrated in Figure 5.12.

We know from our discussion about the auditory pathway that the first place where information from both ears comes together is in the superior olivary nucleus. Briefly, when a sound arrives at one ear it is transduced by the hair cells, elicits firing in the auditory nerve and evokes spikes in the axons that project from the cochlear nuclei to the medial superior olive. The same sound will initiate a similar

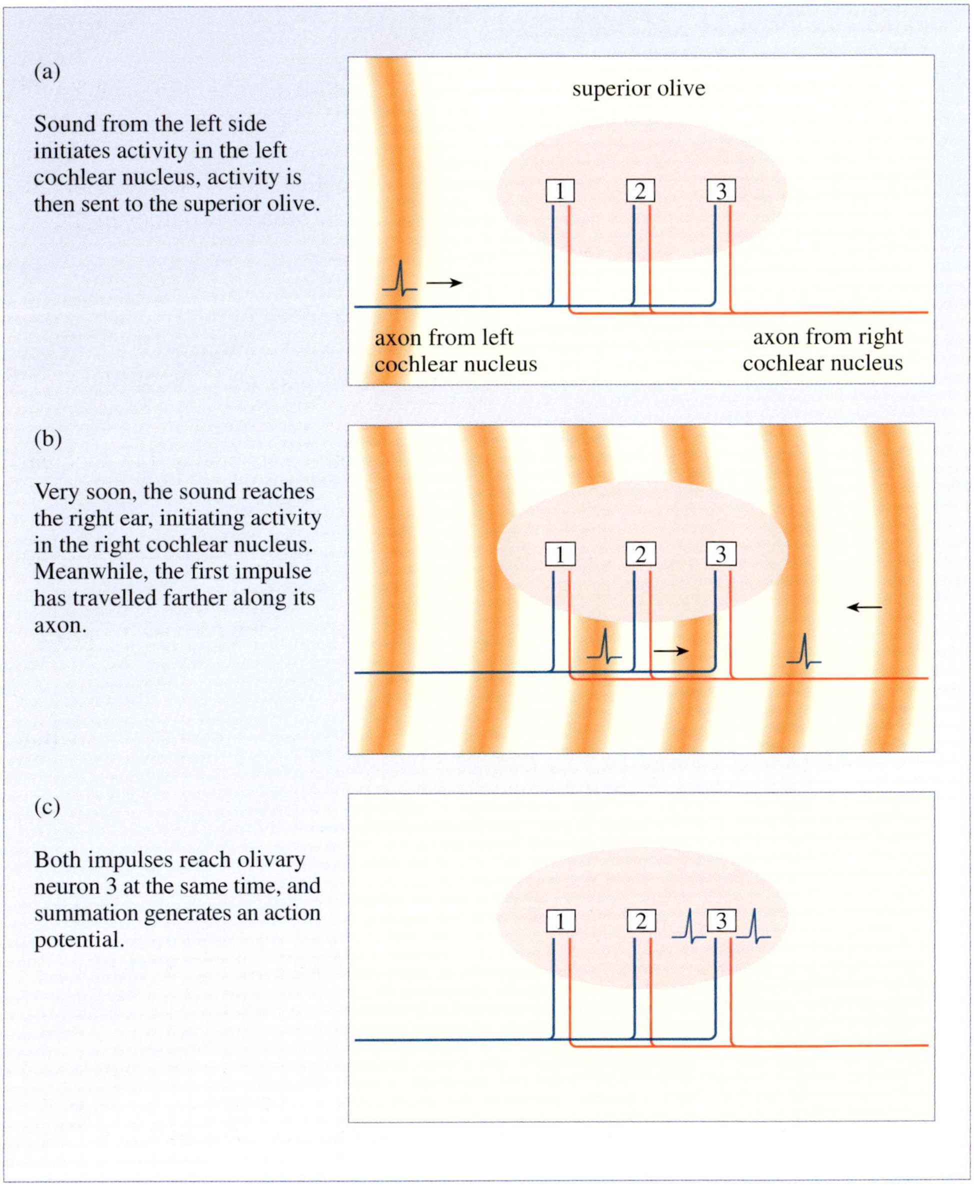

Figure 5.12 Delay lines and coincidence detection.

series of events when it reaches the opposite ear. In Figure 5.12a, the sound reaches the left ear and hence the left cochlear nucleus first resulting in the generation of action potentials that are relayed to cells in the superior olive. A fraction of a millisecond later, sound reaches the right ear, initiating activity in the right cochlear nucleus (Figure 5.12b). By this time however, the impulses travelling from the left cochlear nucleus have travelled further along the axon (which is the delay line). Impulses from both ears reach coincidence at, in this case, olivary neuron 3, which then fires an action potential (Figure 5.12c).

- ❍ According to this illustration, if the sound were directly in front of the person, which of the olivary neurons would fire?
- ● Olivary neuron 2, because the time it would take the action potential to travel from the left and the right cochlear nuclei would be the same and coincidence would occur in the midline.

You can see from this example that the auditory system can extract information about the location of a sound source by attending to which neuron in the superior olivary nucleus fires in response to the sound. This is because neurons will only fire if there is some specific delay between the spikes arriving from the left and right cochlear nuclei, and different neurons fire in response to different delays.

Interaural time delays: continuous tones

Coincidence detectors and delay lines cannot be used to localize a continuous tone.

- ❍ Why?
- ● Because, a continuous tone is always present at both ears and if we don't hear the onset of a sound then our auditory system cannot determine the initial difference in arrival times at the two ears.

So, in order to localize a continuous tone the auditory system uses another kind of temporal information: the time at which the same phase of the sound wave reaches the ear.

Recall that neurons are capable of phase locking to a sound stimulus: they fire at characteristic points or phase angles along the sound wave. A neuron tuned to one frequency would tend to fire, for example when the wave is at baseline (0 degrees), although it may not fire every time the wave reaches this position. A neuron tuned to a different frequency will tend to fire at a different phase angle, such as when the wave is cresting (90 degrees). In both ears impulses produced by neurons tuned to the same frequency will lock to the same phase angle. But, depending when the signals reach the ears, the train of impulses generated in one ear may be delayed relative to the impulse train generated in the other ear.

Imagine you are exposed to a 400 Hz sound coming from the right (Figure 5.13a). At this frequency, one cycle of sound covers about 85 cm, which is more than the 20 cm distance between your ears. After the peak in sound wave passes the right ear, you must wait 0.6 ms, the time it takes the sound to travel 20 cm, before detecting the *same* peak in your left ear. Because the wavelength of the sound wave is much longer than the distance between your ears (85 cm versus 20 cm) you can reliably use the interaural delay in peaks in the wave to determine sound location. What about wavelengths that are shorter than the distance between the ears?

Continuous tones of frequencies above about 1500 Hz produce what are known as **phase ambiguities**. This is because a sinusoidal wave of 1500 Hz has a wavelength about equal to the width of the head. You can see in Figure 5.13b, that both ears will detect a peak in the sound wave at the same time. Clearly, the peaks detected at the ears are different (labelled 1 and 2 in the figure) but as far as the brain is concerned, there would be no phase difference and the sound would be perceived as coming from the front. Head movements may resolve this ambiguity to some extent. However when the wavelength is less than the path difference between the two ears, ambiguities increase; the same phase difference could be produced by a number of different source locations. Phase differences therefore only produce useful cues for frequencies below about 1500 Hz.

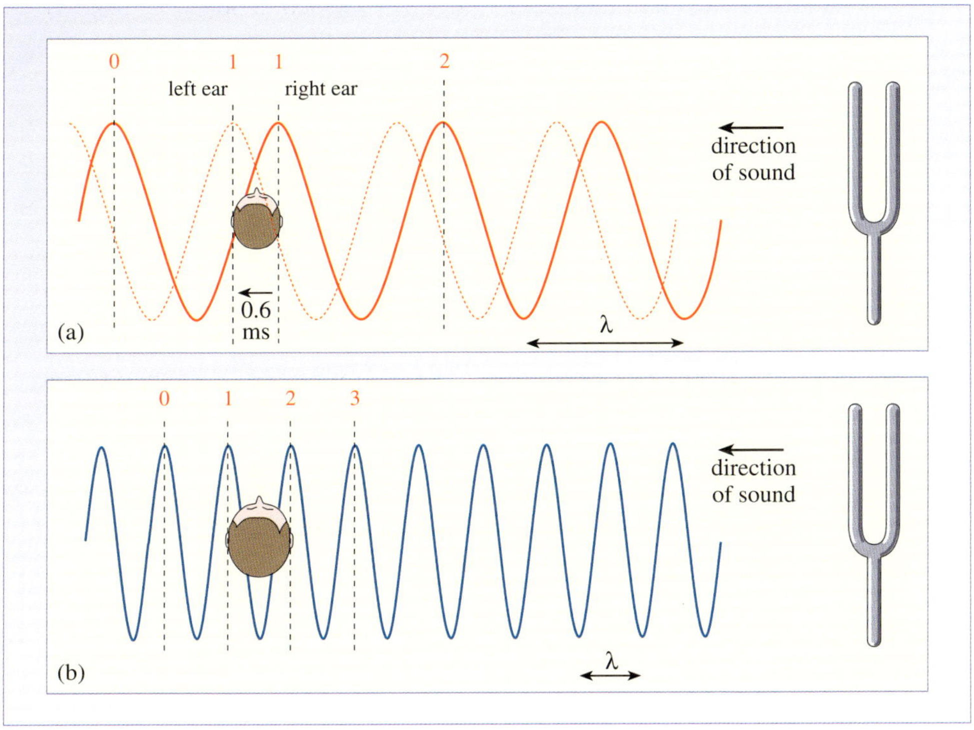

Figure 5.13 Interaural time differences and phase ambiguity. (a) The signal comes from the right and waveform features such as the peak numbered 1 arrive at the right ear (solid line) 0.6 ms before arriving at the left (dotted line). Because the wavelength is more than twice the head diameter, no confusion is caused by the other peaks in the waveform (peaks 0 and 2) and the signal is correctly perceived as coming from the right. (b) The signal again comes from the right but the wavelength is shorter than the head diameter. As a result every feature of cycle 2 arriving at the right ear has a corresponding feature from cycle 1 at the left ear. The listener mistakenly concludes that the source is directly in front.

In fact, even with sounds of frequency as low as 750 Hz we can be confused. Look at Figure 5.13b and note that a 1500 Hz tone will result in a wave-crest appearing at each ear simultaneously. A 750 Hz tone has half that frequency, hence twice the wavelength; crest to crest would cover two head widths. In other words, on a person's head one ear would be at a crest when the other was at a trough of the wave. The brain has no information to indicate which feature is leading, the trough or the crest. Depending on which is ahead, the sound is either coming from the right, or the complete opposite, the left! The problem is easily solved by turning the head from side to side. The ear which was lagging will 'catch up' with the leading ear as it turns towards the sound source. This signals which ear was leading, and hence the sound direction. Head movement can resolve ambiguities effectively until the sound wavelength is equal to the head width (at about 1500 Hz) but beyond this phase differences cannot be interpreted effectively. Nevertheless, we are able to localize high frequency sounds, so there must be another way of doing so.

Interaural intensity differences

The brain has another process for localizing high-frequency sounds (above 1500 Hz): interaural intensity differences.

- ❍ Where does processing of interaural intensity differences take place?
- ● In the lateral superior olivary nucleus.

For any sound, there is a direct relationship between the direction that the sound comes from and the extent to which the intensity of the sound at the two ears differs. If the sound comes directly from the right, the sound will be lower in intensity in the left ear, if it comes from directly in front, the intensity at the two ears is the same and with sound coming from intermediate directions, there are intermediate intensity differences. Intensity differences between the ears can result from two factors: differences in the distance the sound must travel to the two ears and differences in the degree to which the head casts a **sound shadow**. The greater the sound shadow cast by the head, the greater the level difference between the ears. The extent of the sound shadow cast by the head depends on the frequency of the sound. Low-frequency sounds have a wavelength that is long compared to the size of the head. The sound therefore bends very well around the head and there is very little sound shadow cast. In contrast, high-frequency sounds have a wavelength that is short compared to the dimensions of the head. This means that the head casts a significant sound shadow.

- ❍ What would be the consequences of the difference in sound shadow cast by the head in response to low-frequency and high-frequency sounds?
- ● For low frequencies, the sound shadow cast by the head would cause a minimal difference in the intensity of sound at the two ears and so there would be little scope for using interaural intensity differences in order to localize the sound. For high frequencies, the sound shadow cast by the head would cause a

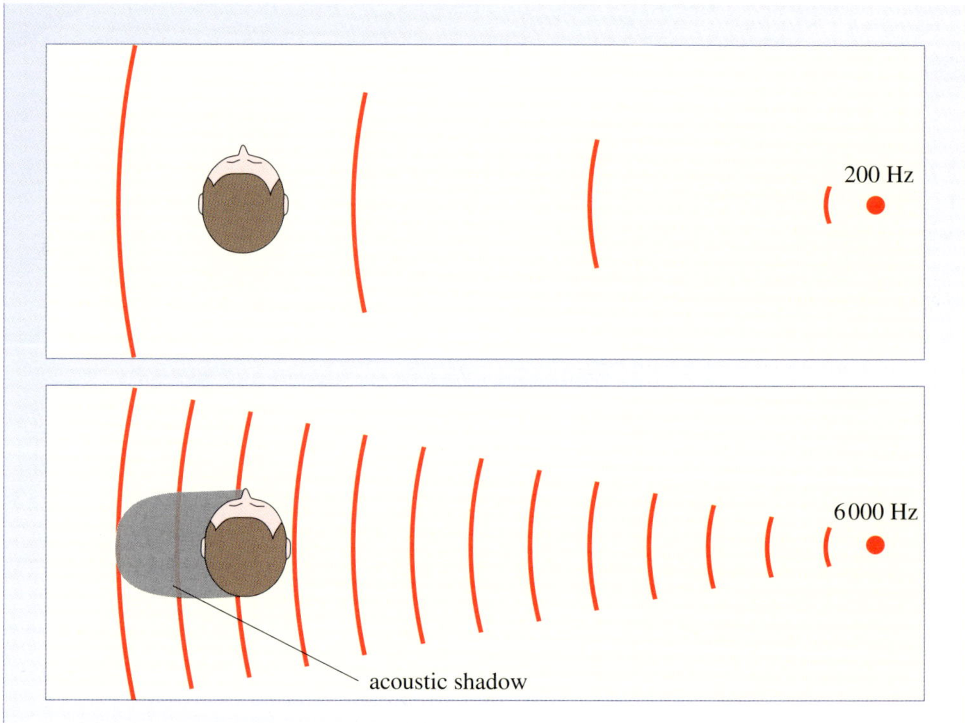

Figure 5.14 Low-frequency tones are not affected by the listener's head, so the intensity of a 200 Hz tone is the same at both ears. High-frequency tones (e.g. 6000 Hz) are affected by the presence of the listener's head and result in an acoustic shadow that decreases the intensity of the tone reaching the listener's far ear.

significant difference in the intensity of the sound at the two ears and so facilitate the use of interaural intensity differences.

In fact, interaural differences in intensity are negligible at low frequencies, but may be as large as 20 dB at high frequencies (Figure 5.14).

5.4.2 Localization of sound in the vertical plane

Much of our ability to localize sound in the vertical plane is due to the shape of the outer ear, in particular the pinna. The pinnae provide a monaural cue to localization. The bumps and ridges on the pinnae produce reflections, and delays between the direct path and the reflected path make vertical localization possible. Vertical localization is seriously impaired if the convolutions of the pinnae are covered.

5.4.3 Distance cues

There are two main cues available that allow us to judge the distance to a sound source. The first of these is the sound pressure level. Sound pressure level drops by 6 dB each time the distance that a sound travels doubles. In other words, if the sound pressure level of a sound is 60 dB SPL when its source is 1 m from you, then it will be 54 dB SPL if you move back another metre so that you are now 2 m away from its source. Therefore lower sound pressure levels indicate a greater distance. A second cue relates to the frequency of a sound. When a sound travels over a distance, the high-frequency components attenuate to a greater extent than the low-frequency components. This means that sounds that are further away tend to be richer in low frequencies and therefore have a lower pitch.

You should now read Chapter 5 of the Reader, *Hearing the world* by Chris Darwin. At this point you should also read Chapter 6, *Hearing impairments: causes, effects and rehabilitation* by David Baguley and Don McFerran. The topics in this chapter are not covered in the block text but are nonetheless an integral part of the course and should be read carefully.

Summary of Section 5.4

For precise localization of a sound source, binaural cues are required.

Two types of binaural cue are used to localize non-continuous sounds in the horizontal plane: interaural time differences, which are most efficient for low-frequency sounds (20–1500 Hz) and interaural intensity cues, which are important for high-frequency sounds (1500–20 000 Hz). The frequency responses in the superior olive reflect these differences. The medial superior olive includes neurons that are responsive to low-frequency inputs, while the cells of the lateral superior olive are most sensitive to high-frequency stimuli.

The mechanism involved for the detection of interaural time differences is believed to involve a series of delay lines and coincidence detectors.

For localization of continuous tones, interaural phase differences are used.

Information about the location of a signal in the vertical plane is provided by the pinnae.

We are able to judge the distance to a sound source using cues related to the decay of the signal with distance. Both the SPL and the spectral components of a signal are dependent on the distance between the signal and the listener.

Question 5.1

(a) If two tones are broadcast through headphones at an intensity of 100 dB SPL, which will sound louder, a 150 Hz tone or a 1000 Hz tone? Why?

(b) How intense must a 100 Hz tone and a 1000 Hz tone be (in dB SPL) in order to have a loudness level of 50 phons?

Question 5.2

(a) How are beats generated? (b) How are they perceived?

Question 5.3

Why is it that when you play music softly that has been recorded at a fairly high level, you cannot hear the very high and very low frequencies?

Question 5.4

How does the use of masking experiments support the place code hypothesis for pitch perception?

Question 5.5

Why are interaural time differences not very useful for localizing high-frequency sounds?

Question 5.6

What properties of a sound determine its pitch and its timbre?

STUDY FILE

Activity 5.2 Sound localization

There are two activities associated with sound localization, which can be found on the CD-ROM. Details of both these activities are given in the Block 3 *Study File*.

Balance: the vestibular system

6

Although a primary function of the inner ear is to perceive sound, it has other, equally important functions: to help the body maintain its postural equilibrium, and to coordinate the position of the head and the movement of the eyes. The apparatus that performs these tasks is known as the **vestibular system** (Figure 6.1), and consists of the vestibule and semicircular canals. The vestibule houses two membranous sacs, the **utricle** and the **saccule**, which are known as **otolithic organs**. They respond to changes in the position of the head with respect to gravity (*linear acceleration*). The information that these organs deliver is **proprioceptive** in character, dealing with events within the body itself, rather than **exteroceptive**, dealing with events outside the body (as in the case of the response of the cochlea to sound). There are three semicircular canals, each oriented at right angles to one another which respond to rotational movements (*angular acceleration*).

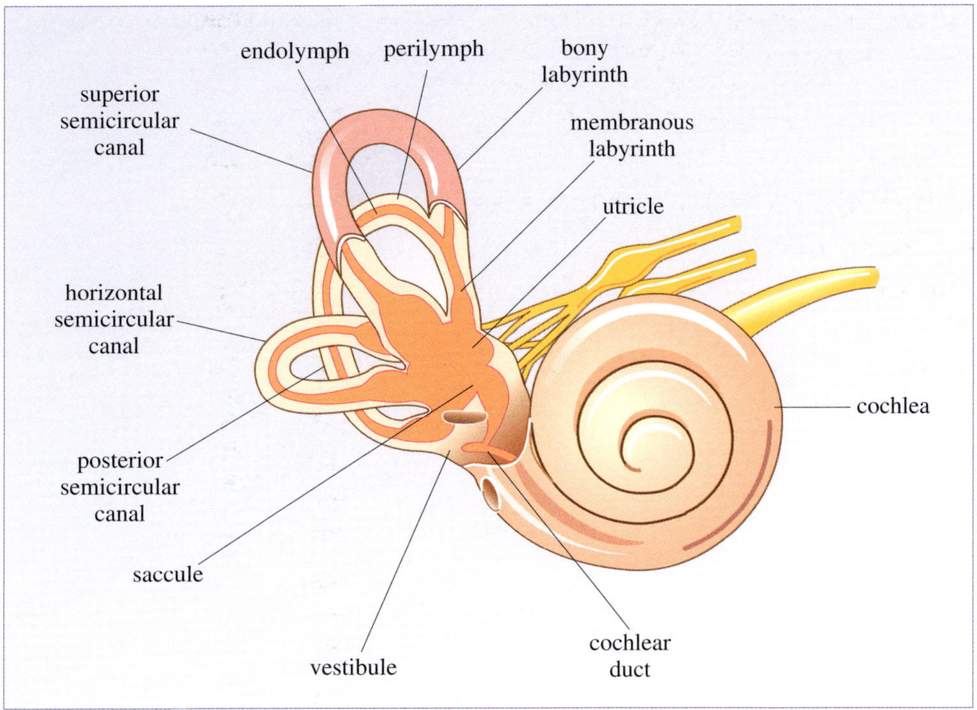

Figure 6.1 The vestibular system.

6.1 The utricle and the saccule: linear acceleration

The utricle and the saccule contain the gravity receptors and also respond to linear acceleration of the head. They signal linear acceleration and head tilt relative to earth's gravity, and allow you to sense the vertical acceleration of a lift or a downward and upward jump. They also sense the horizontal acceleration you feel when you step on the car's accelerator.

The receptors called **maculae** (Figure 6.2 overleaf) are composed of a single patch of sensory cells attached to the inner surface of the utricle and saccule. In the utricle, the macula projects from the anterior wall and lies primarily in the horizontal plane. It therefore detects acceleration in the horizontal plane. In the saccule, the macula is in the vertical plane and has an elongated shape resembling the letter 'J'. It detects movement in the sagittal plane (up and down, forward and back). The sensory cells are hair cells which, as in the cochlea, have fine hairs (nonmotile stereocilia and a single motile kinocilium) projecting from their apical ends. Covering the entire macula is a delicate acellular structure, the **otolithic membrane**. The surface of the membrane is covered by a blanket of calcium carbonate crystals, called **otoliths**.

When the head undergoes linear acceleration, the movement of the otolithic mass lags behind that of the head because of its inertia. The movement of this mass is communicated to the otolithic membrane, which then shifts with respect to the underlying epithelium. This motion in turn deflects the hair bundles that link the membrane to the macula, thus eliciting an electrical response in the hair cells. Recall that hair cells are stimulated (depolarized) when the stereocilia are deflected towards the tall edge and are inhibited (hyperpolarized) when they are deflected in the opposite direction. Hair cells in the maculae are oriented in a number of different directions. So when the head moves and deflection of the hair bundles occurs, some hair cells will be excited and some will be inhibited (Figure 6.2). This means that the hair cell response will provide the central nervous system with a unique pattern of signals for any linear acceleration of any magnitude and orientation.

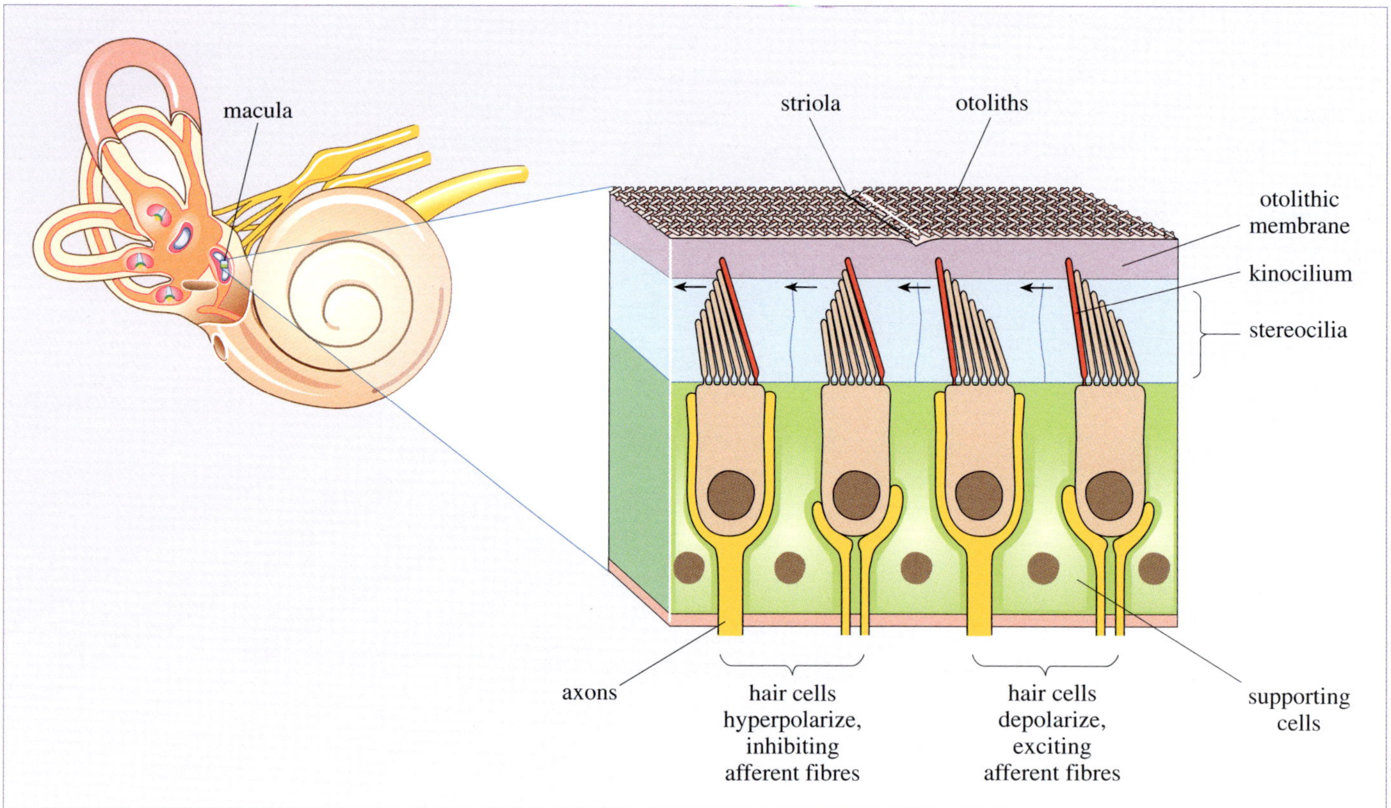

Figure 6.2 Diagram of a macula. Hair cells have hair bundles that project into the otolithic membrane, a gelatinous mat embedded with calcium carbonate stones (otoliths). Not all hair cells are oriented in the same direction. The response of an individual hair cell to the tilt of the head depends on the direction in which the hairs are bent by the gravitational force on the otoliths. When the head is tilted in one direction, some hair cells will depolarize and excite the afferent fibre. When the head is tilted in the opposite direction these hair cells will hyperpolarize and inhibit firing in the afferent fibre. Other hair cells exhibit the opposite behaviour.

6.2 The semicircular canals: angular acceleration

The three semicircular canals respond to angular acceleration. Angular acceleration occurs whenever an object alters its rate of rotation about an axis. Our head therefore undergoes angular acceleration during turning or tilting motions of the head, rotatory body movements and turning movements during passive or active locomotion. The three semicircular canals in each ear detect these angular accelerations and report their magnitudes and orientations to the brain.

Because the semicircular canals are oriented at right angles to one another they are able to detect movements in three-dimensional space. Each canal has an expanded end section called an **ampulla**, which opens into the vestibule. The hair cells are located in the ampulla on a saddle-shaped ridge of tissue called a **crista** (Figure 6.3). The cilia of the hair cells project up into a gelatinous mass, the **cupula**.

Like the otolithic organs, the semicircular canals detect accelerations by means of the inertia of the internal material, which is endolymph. Here, however, it is the mass of the endolymph itself that responds to acceleration. Consider a simple situation in which there is a smoothly increasing rotatory motion, and hence a constant angular acceleration, about an axis passing perpendicularly through the centre of a semicircular canal. As the head rotates faster and faster, the endolymph, because of its inertia, tends to lag behind and therefore rotates within the semicircular canal in a direction opposite to that of the head. The motion of the endolymph can be illustrated using a cup of water and a small piece of paper (about the size of a pea) placed on the surface of the water somewhere around the edge. While gently rotating the cup around its vertical axis in a clockwise direction, watch the small piece of paper. As the cup begins to turn, the paper tends to maintain its original orientation in space. This means that the water must rotate counterclockwise in the cup (otherwise the paper would also rotate clockwise). At the conclusion of

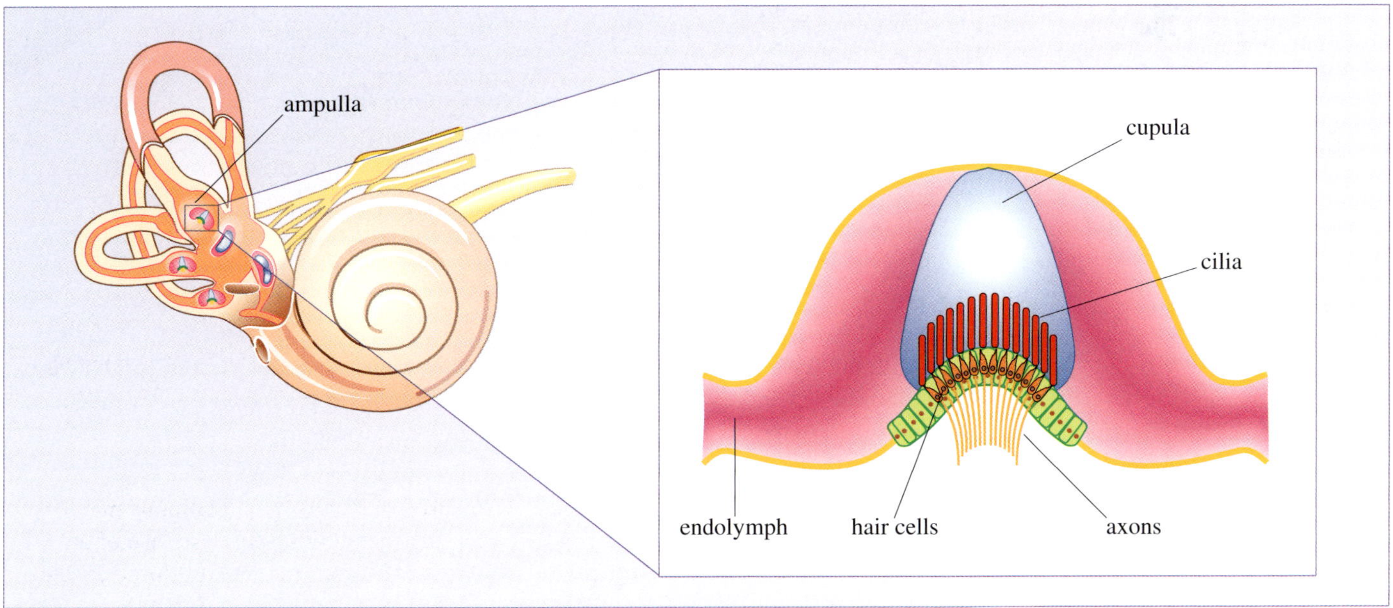

Figure 6.3 Cross-section through the ampulla of a semicircular canal, showing the crista. The cilia of the hair cells penetrate into the gelatinous cupula. The cupula is bathed in endolymph, which fills the canals.

the turning motion, when the cup decelerates and then stops, the paper begins to move in a clockwise direction. This is essentially what the endolymph does in the semicircular canals. However the fluid cannot move freely around the canal because the cupula of each crista extends across the canal at its widest region (the ampulla). So, when the endolymph moves, it pushes against the surface of the cupula causing it to bend, and resulting in the stimulation of the hair cells in the crista. The process by which the semicircular canals signal angular acceleration is shown in Figure 6.4.

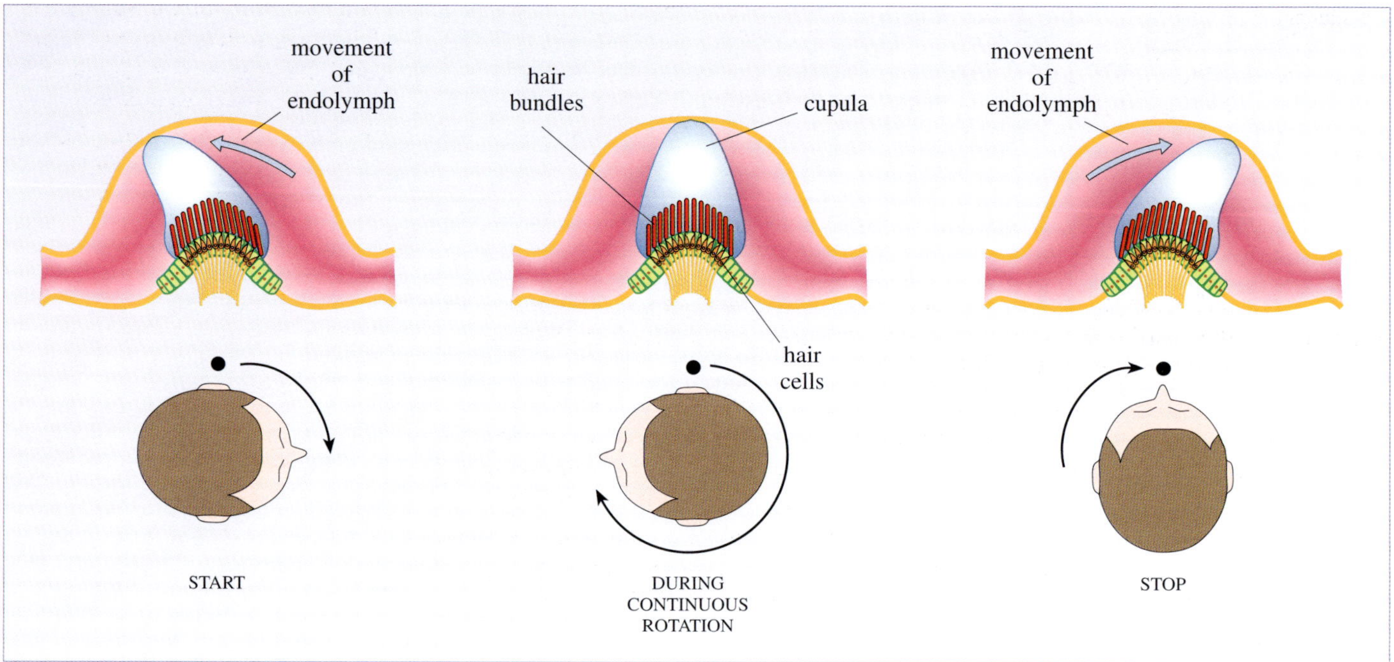

Figure 6.4 Schematic diagram of the movement of the cupula in response to rotatory movement of the head. During continuous rotation, the bending of the cupula stops after 15–20 seconds.

When the head starts turning, the fluid (endolymph) in the semicircular canals lags behind at first, because of its inertia, and consequently pushes against the cupula. This distorts it, and consequently bends the hair tufts of the crista inside. Because the fluid is lagging, the hairs are bent in the opposite direction to the head movement. As the movement of the head continues, the endolymph and the cupula move at the same rate as the rotatory movement of the head and the cilia become erect. As long as the head is moving in a constant direction at a constant speed, or is rotating with a constant angular speed, the cupula and cilia remain erect (as is the case with no head movement). When head movement stops, the inertial force of the endolymph carries the cupula forward and the cilia are bent in the *forward* direction. Therefore, starts and stops bend the cilia whereas constant motion produces no deflection. All the hair bundles in each of the semicircular canals share a common orientation. Therefore acceleration in one direction depolarizes hair cells and excites afferent neurons while acceleration in the opposite direction hyperpolarizes the hair cells and diminishes spontaneous neural activity.

Naturally, the arrangement of semicircular canals is mirrored on both sides of the head. This means that generally, when the cells on one side of the head are excited, the cells on the other side of the head will be inhibited. This is illustrated in Figure 6.5.

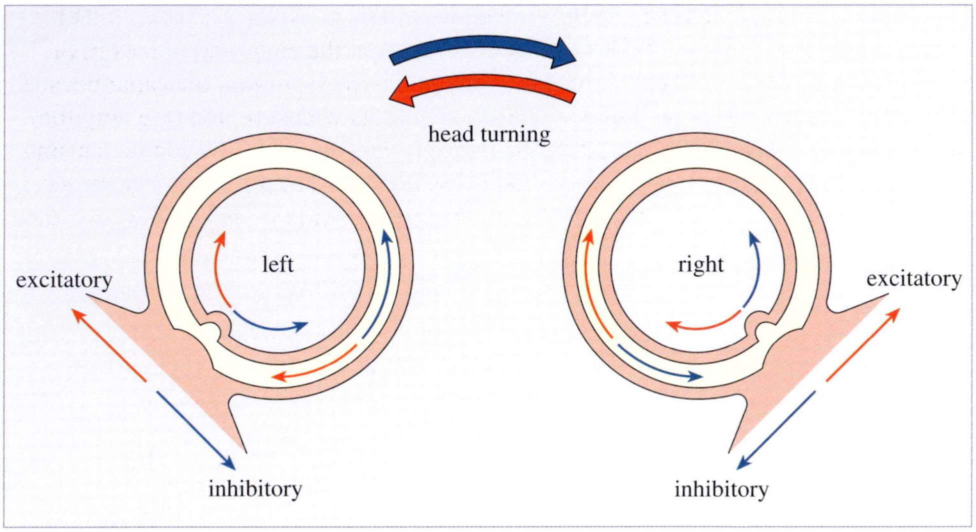

Figure 6.5 View of the horizontal semicircular canals from above, showing how the canals on each side of the head work together to signal head movement. Because of inertia, rotation of the head in a counter-clockwise direction causes the endolymph to move clockwise with respect to the canals. This deflects the stereocilia in the left canal in an excitatory direction, thereby exciting afferent fibres on this side. In the right canal, hair cells are hyperpolarized and afferent firing there decreases.

The hair cells of the vestibular system send information about head acceleration to the brain via the vestibular component of the VIIIth cranial nerve. Information travels via the **vestibular nerve** to the **vestibular nuclei** in the medulla which then distribute it to centres further up the pathway. This network of vestibular connections is responsible for various reflexes that the body uses to compensate for head movement and the perception of movement in space. The **vestibulo-ocular reflexes** are responsible for keeping the eyes stationary with respect to the outside world when the head moves. If your eyes moved with your head, the image on the retina would also move and cause blurring. This is avoided by the vestibular-ocular reflex causing the eye muscles to produce just the right amount of opposite eye movement, to nullify the head movement.

This can be demonstrated by shaking your head while reading this paragraph. You will find that you have no trouble staying focused on this page. However, if instead you shake the book from side to side you will find that the eyes cannot keep up with the movement, so the writing blurs.

Why would shaking your head and shaking the book result in these different effects?

When you move your head and keep the book still, the vestibular system signals to the brain how fast the head is moving and the **oculomotor system** then uses this information to stabilize the eyes in order to keep the visual image motionless on the retina. When you move the book but keep your head still, vision is the only cue the brain has to stabilize the image and since visual processing is much slower than vestibular processing for image stabilization, the eyes are not able to keep up with the motion of the book.

There are three different vestibulo-ocular reflexes, which arise from the three major components of the vestibular system:

1 the rotational vestibulo-ocular reflex which compensates for head rotation and receives input primarily from the semi-circular canals;

2 the translational vestibulo-ocular reflex which compensates for linear head movements;

3 the ocular counter-rolling response which compensates for head tilt in the vertical.

The second and third reflexes receive their input predominantly from the otolithic organs and are sometimes called otolith reflexes.

Of the three reflexes, the rotational reflex is the simplest. When the semicircular canals sense head rotation in one direction, the eyes slowly rotate in the opposite direction. In trying to compensate the movement during sustained head rotation, the eyes would eventually be rotated as far as they could go. This situation is avoided, because the eyes rotate only so far, then make a rapid resetting movement, back to the central position. They then start to make the compensatory rotation again.

You should now read Chapter 7 of the Reader, *The vestibular system* by David Furness. At this point you may also like to read Chapter 28, *Interaction between the senses: vision and the vestibular system* by Rollin Stott. It provides a more detailed description of the vestibulo-ocular reflex as well as a discussion of how the vestibular system contributes to postural stability. The question of what causes motion sickness is also discussed. Chapter 28 will be revisited in Block 7.

Summary of Section 6

The vestibular system functions to help the body maintain its postural equilibrium and to coordinate the position of the head and the movement of the eyes.

The utricle and the saccule, sometimes known as the otolithic organs, detect the static position of the head and also respond to linear acceleration of the head. The hair cells are located on a flat plaque called a macula.

The three semicircular canals respond to angular acceleration. Angular acceleration occurs whenever an object alters its rate of rotation about an axis. Hair cells that respond to changes in angular acceleration of the head are located on a ridge called a crista.

Question 6.1

Which part of the vestibular system would be activated by linear acceleration of the head? How would information about linear acceleration be encoded?

Objectives for Block 3

Now that you have completed this block and Chapters 1 to 7 of the Reader you should be able to:

1 Define and use, or recognize definitions and applications of, each of the terms printed in **bold** in the text.

2 Describe the nature of sound and the basic characteristics of simple sounds. *(Questions 2.1, 2.2, 2.3, 2.4 and 5.2)*

3 Distinguish between the major anatomical components of the outer, middle and inner ear. *(Questions 3.1 and 3.3)*

4 Describe the function of the outer, middle and inner ear in hearing. *(Questions 3.1 and 3.2)*

5 Describe the structure of the cochlea. *(Question 3.3)*

6 (a) Describe the excitation process: the travelling wave pattern of vibration on the basilar membrane and how this vibration leads to depolarization of the hair cells and stimulation of auditory nerve fibres. (b) Explain the structural arrangements in the organ of Corti that make this possible. *(Questions 3.3, 3.4 and 3.5)*

7 (a) Explain the difference between the four coding mechanisms used in order to transmit information from the ear to the brain. (b) Explain what is meant by: threshold tuning curve, characteristic frequency, firing rate, place theory of hearing, phase locking, volley theory of hearing. *(Questions 4.1, 4.2, 4.3, 4.4 and 4.5)*

8 Describe the ascending auditory pathway and the function of the main nuclei involved. *(Question 4.6)*

9 Describe how brain imaging has contributed to our understanding of the brain areas involved in hearing.

10 Describe the basic principles of psychophysics and explain what is meant by differential sensitivity, difference limen, just noticeable difference and Weber's law. *(Questions 5.1 and 5.3)*

11 Explain the difference between intensity and loudness and between frequency and pitch. *(Questions 5.4 and 5.6)*

12 Explain the concept of masking and how it contributes to our understanding of the place theory of frequency coding. *(Question 5.4)*

13 Describe the use of intensity and timing cues in sound localization. *(Question 5.5)*

14 Describe the main causes of hearing impairments and the methods used to rehabilitate hearing-impaired individuals.

15 Describe the main components of the vestibular system and how they contribute to our ability to maintain our balance. *(Question 6.1)*

Answers to questions

Question 2.1

The speed of sound in air at 293 K is 343 m s^{-1} and in water it is 1480 m s^{-1} (see Table 2.1). The wavelength can be found from:

$$\lambda = c/f$$

where c = speed, f = frequency and λ = wavelength. Substituting in numbers gives:

in air: $\lambda = 343\ \text{m s}^{-1} / 262\ \text{Hz} = 1.3\ \text{m}$

in water: $\lambda = 1480\ \text{m s}^{-1} / 262\ \text{Hz} = 5.6\ \text{m}$

Question 2.2

It allows us to work with a much more convenient range of numbers to describe the extremely wide dynamic range of the auditory system.

Question 2.3

From equation 2.2 we have:

$$\text{relative intensity} = 10 \log_{10}(I_1/I_2)\ \text{dB}$$

If $I_1/I_2 = 2$, then the relative intensity in dB is $10 \log_{10} 2 = 3.01$ dB. This is a useful figure to remember – a drop of 3 dB corresponds to a halving of the intensity.

Question 2.4

The notch filter will attenuate frequencies in the notch region and leave higher and lower frequencies unaffected. Your sketches should therefore look like Figure 2.11.

Question 3.1

The first way in which the middle ear enhances the efficiency of sound transfer is to do with the relative sizes of the tympanic membrane and the stapes footplate (which is connected to the oval window). Measurements have shown that the area of the tympanic membrane that vibrates in response to high intensity sound is 55 mm^2.

The stapes footplate which makes contact with the oval window has an area of about 3.2 mm^2, which is considerably less than the effective area of the tympanic membrane. So, if all the force exerted on the tympanic membrane is transferred to the stapes footplate, then the force per unit area must be greater at the footplate because it is smaller than the tympanic membrane. Put simply, if the same force is applied to a large area and to a small area, the force applied to the smaller area will result in a bigger pressure change. You know that if you hit a wall with a hammer you make a small dent but if you hit a nail with a hammer swung with the same force, all the force is concentrated on a small point and the nail is driven into the wall. In fact the tympanic membrane and the footplate differ in size by a factor of 17 (55 mm^2 / 3.2 mm^2 = 17) so pressure at the footplate (force per unit area) is 17 times greater than at the tympanic membrane and therefore the air pressure can stimulate the fluid filled inner ear.

The second way in which the middle ear enhances the efficiency of sound transfer is through the lever action of the ossicles. Figure 3.2 shows how a lever system can increase the force of an incoming signal. In the figure, the lever is pivoting around a fulcrum at point C. The distance D_1 between the fulcrum and the point of the applied force is larger than the distance D_2, between the fulcrum and the position of the resultant force. The increase in force due to lever action is given by the formula:

$$F_{\text{resultant}} = F_{\text{applied}} \times (D_1/D_2)$$

Therefore the closer the fulcrum is to the point of the resultant force, the larger this force will be. The ossicles of the middle ear are arranged so that they act like a lever. The length of the malleus corresponds to D_1 (the distance between the applied force and the fulcrum), while the incus acts as the lever portion between the fulcrum and

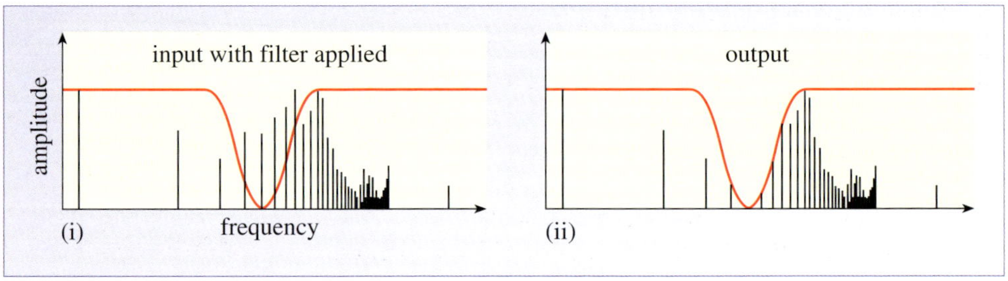

Figure 2.11 The notch filter cuts off frequencies in a certain range but allows frequencies below and above the range to pass through the filter.

the resultant signal (D_2). Measurements of the length of these two bones indicate that the lever system of the ossicles increases the force at the tympanic membrane by a factor of 1.2 at the stapes. In addition, the tympanic membrane tends to buckle as it moves causing the malleus to move with about twice the force. So, overall the increase in pressure at the stapes footplate is in the region of $17 \times 1.2 \times 2 = 40.8$. The reduction in sound level caused by the fluid/air interface is estimated to be about 30 dB. Therefore the middle ear counteracts this reduction.

Question 3.2

pinna → external auditory canal → tympanic membrane → malleus → incus → stapes → oval window → scala vestibuli → helicotrema → scala tympani → round window

Question 3.3

The cochlea has a spiral shape resembling the shell of a snail. Unravelled, the cochlea's hollow tube is about 32 mm long and 2 mm in diameter. The tube of the cochlea is divided into three chambers: the scala vestibuli, the scala media (or cochlear duct) and the scala tympani. The three scalae wrap around inside the cochlea like a spiral staircase. The scala vestibuli forms the upper chamber and at the base of this chamber is the oval window. The lowermost of the three chambers is the scala tympani. It too has a basal aperture, the round window, which is closed by an elastic membrane. The scala media or cochlear duct separates the other two chambers along most of their length. The start of the cochlea, where the oval and round windows are located is known as the basal end, while the other end, the inner tip, is known as the apical end. The scala vestibuli and the scala tympani communicate with one another via the helicotrema, an opening in the cochlear duct at the apex. Both scala vestibuli and scala tympani are filled with the same fluid known as perilymph while the scala media is filled with endolymph.

Between the scala vestibuli and the scala media is a membrane called Reissner's membrane and between the scala tympani and the scala media is the basilar membrane. Lying on top of the basilar membrane within the cochlear duct is the organ of Corti, and hanging over the organ of Corti is the tectorial membrane. In response to sound entering the cochlea, the fluid within the cochlea vibrates. The key factor in the response of the inner ear is the mechanical response of the basilar membrane and organ of Corti. These two structures translate the mechanical vibrations of the inner ear fluids into neural responses in the auditory nerve. The vibration of the fluids causes the basilar membrane to move which creates a shearing motion between the basilar membrane and the overlying tectorial membrane. This in turn causes the cilia of the hair cells contained within the organ of Corti to bend. The bending of the cilia results in the nerve fibre at the base of the hair cell initiating a neural potential that is sent along the auditory nerve in the form of action potentials. Thus the hair cells in conjuction with the basilar membrane translate mechanical information into neural information.

Question 3.4

It is the motion set up on the basilar membrane in response to movement of the cochlea fluids. The wave propagates from the base of the membrane towards the apex. The point of maximal displacement of the wave is determined by the frequency of the incoming sound.

Question 3.5

Endolymph is found in the scala media and perilymph is found in the scala vestibuli and scala tympani. Endolymph has an ionic concentration similar to that of intracellular fluid, high K^+ and low Na^+ (even though it is extracellular). Perilymph has an ionic content similar to that of cerebrospinal fluid, low K^+ and high Na^+. Because of the ionic concentration differences, the endolymph has an electrical potential that is about 80 mV more positive than the perilymph. The stereocilia of the hair cells are bathed in endolymph while the base of the hair cells and their afferent dendrites are bathed in perilymph. When the stereocilia are bent by movement of the basilar membrane, they either depolarize or hyperpolarize, depending on the direction in which they are bent. The receptor potential, which is either above or below the resting potential of the hair cell, results from opening or closing potassium channels in the tips of the stereocilia. When the cell depolarizes, K^+ channels open and more K^+ enters the cell. When the cell hyperpolarizes, K^+ channels, which are normally partially open, close and inward movement of K^+ is prevented.

Question 4.1

Phase locking is the consistent firing of an auditory neuron at the same phase of each cycle of a sound wave. At low frequencies, the neuron will fire action potentials at some constant location on the wave (peak or trough, for example) so that the frequency of the sound can be determined from the frequency of the neuron's action potentials. For higher-frequency sounds, neurons may not fire on every cycle even though they fire at the same point on the cycle. A group of such neurons can encode the frequency of the sound wave if their activity is pooled.

Question 4.2

Recall that low frequencies create a rather broad or flat pattern of vibratory activity on the basilar membrane – nearly the entire membrane moves although the peak in the wave is towards the base. High frequencies on the other hand create a wave that is very localized – only a small part of the membrane moves. This means that for low frequencies, the displacement pattern on the membrane is much less specific and localizable than the peak displacement at high frequencies.

Question 4.3

(a) The travelling wave on the basilar membrane has a peak amplitude at a location determined by the frequency of the incoming sound wave. For low-frequency sounds the peak is located at the apical end of the membrane while for high-frequency sounds the peak is located at the basal end. Hair cells are located along the basilar membrane. Cells that are located at the position where the wave peaks are stimulated, resulting in the generation of action potentials in the auditory nerve fibres contacting those hair cells. Fibres on the outside of the auditory nerve innervate the basal hair cells and therefore fire in response to high-frequency sounds whereas fibres on the inside of the auditory nerve innervate apical hair cells and therefore fire in response to low-frequency sounds. So when a sound of a certain frequency stimulates the basilar membrane, the brain receives information about the frequency of the sound, as a consequence of which the fibres in the auditory nerve fire action potentials at the highest relative rate.

(b) The higher the intensity of the sound impinging on the ear, the greater the amplitude of the wave produced on the basilar membrane. The higher amplitude wave causes nerve fibres to fire at a greater rate (the number of action potentials per second is higher for a high-amplitude wave compared to a low-amplitude wave) or causes more neurons to fire (auditory fibres have different thresholds and the greater the displacement of the membrane the greater the number of neurons that reach threshold).

Question 4.4

Frequency is determined by the place code (which neurons are firing) and by a temporal code (neurons fire in bursts that phase lock to the stimulus frequency). Intensity is determined by the firing rate (more spikes per burst for louder sounds) and the number of neurons (at high intensity, more spikes are produced from both positions). (see Figure 4.11 opposite).

Question 4.5

To determine the characteristic frequency of an auditory nerve fibre you would construct a tuning curve. Tuning curves indicate the sound pressure level at the eardrum that is just sufficient to elicit a detectable increase in the firing rate of an auditory nerve fibre, as a function of the frequency of a pure tone stimulus. For each fibre the lowest intensity of a pure tone that will produce a detectable response across a range of pure tones is determined. The frequency of the tone for which the threshold of a given fibre is lowest is called the critical or characteristic frequency.

Question 4.6

Binaural processing of information begins in the superior olivary complex, using interaural time delays and interaural intensity differences. In the MSO, neurons increase their firing rate in response to sounds from both ears, and will increase their discharge rate even further when sounds reach both ears with a certain delay (interaural time delays, EE units). In the LSO, neurons increase their firing rate in response to sounds in the ipsilateral ear and are inhibited from firing by sounds in the contralateral ear. So stimulation from both ears may decrease the firing rate of neurons compared to stimulation by one ear (EI units).

Question 5.1

(a) They will sound equally loud – they both fall on approximately the same equal loudness contour (see Figure 5.4).

(b) The 100 Hz tone must be 60 dB SPL and the 1000 Hz tone about 50 dB SPL (Figure 5.4).

Question 5.2

(a) Beats occur when two tones of slightly different frequency are broadcast simultaneously. As the relative phase of two simultaneously applied tones changes continuously, so the tones alternately reinforce and cancel one another.

(b) They are perceived as a single tone with a pitch midway between the two tones but periodically varying in loudness: waxing and waning of loudness. The intensity varies at a rate equal to the frequency difference of the two tones.

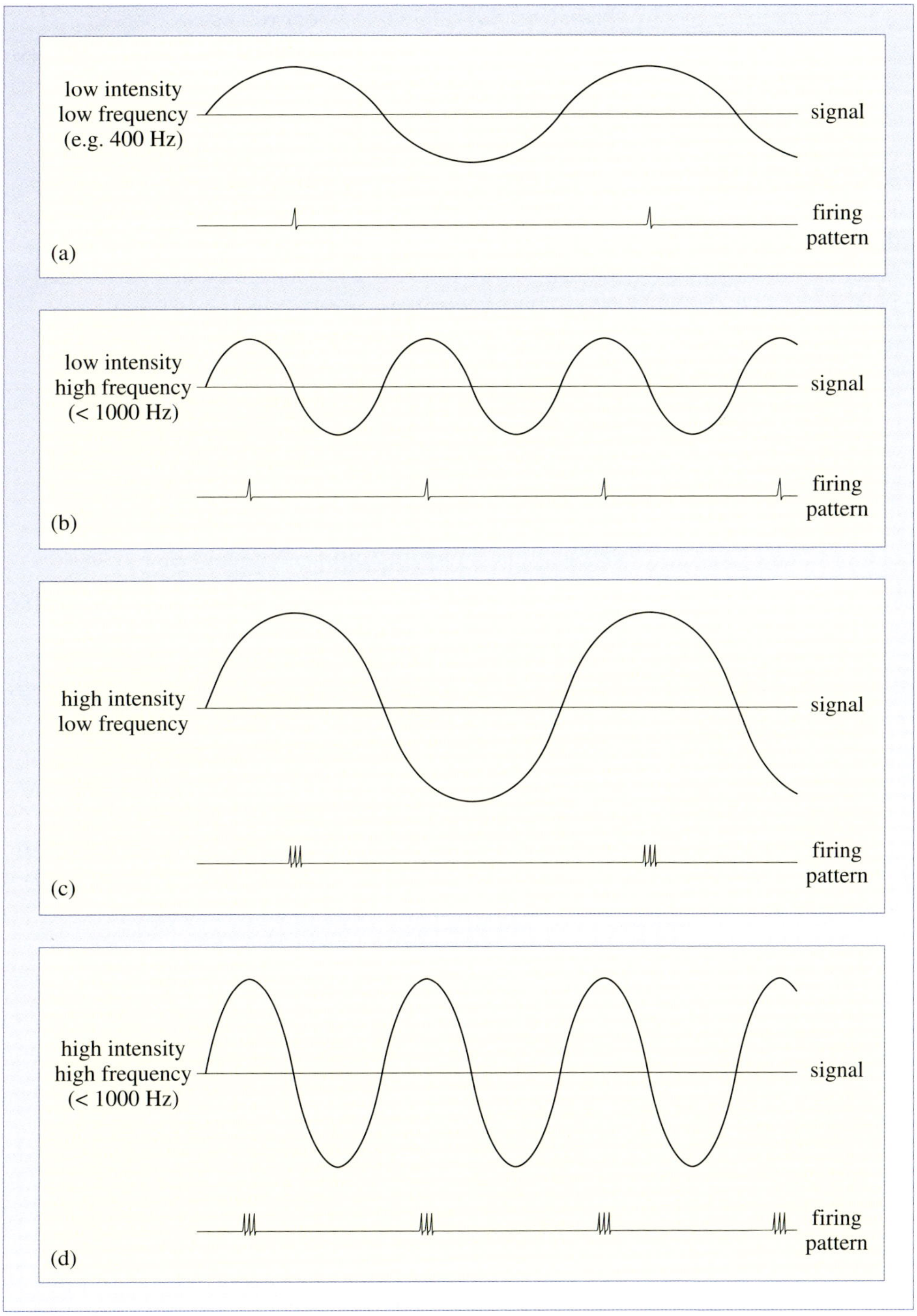

Figure 4.11 Answer to Question 4.4.

Question 5.3

When music is played loudly, above about 80 dB SPL, all tones from about 30 Hz to about 5000 Hz have about the same loudness (they fall on the same loudness contour – see Figure 5.4). However, when you turn the intensity of the music down, all frequencies do not sound equally loud. At 10 dB for example, frequencies below about 400 Hz (the bass notes) and those above about 8000 Hz (the treble notes) are inaudible. So, if you play the music softly, you won't hear the very low and very high frequencies.

Question 5.4

When a test tone is played in the presence of a masking tone or in the presence of masking noise, our ability to hear the test tone is impaired. This is because according to the place code hypothesis there is overlap in the place of vibration on the basilar membrane of the test tone and the masker.

Question 5.5

For high frequencies, the wavelength of the sound is less than the distance between our ears. This means that the delay in the arrival of the sound at the two ears can create phase ambiguities.

Question 5.6

Pitch is mainly determined by the frequencies of the low numbered harmonics in a sound, whereas timbre is determined by which frequency regions have more energy – the relative intensities of different harmonics.

Question 6.1

The parts of the vestibular system that relay information about linear acceleration to the brain are the otolithic organs located in the utricle and saccule. Individual hair cells are oriented in many directions within the utricle and saccule. During linear acceleration, the otoliths lag behind the movement of the head thereby bending some of the hair cell stereocilia and causing excitation and an increase in firing rate of neurons contacting the hair cells. When a constant speed is reached, the otoliths 'catch up' to the speed of the head, allowing the stereocilia to straighten, and the firing rate of neurons returns to resting rate. Upon deceleration, the head slows down and the otoliths continue in the original direction, once again lagging and bending the stereocilia in the opposite direction. This causes inhibition. For other hair cells, oriented in the opposite direction, the initial acceleration causes inhibition, and deceleration causes excitation. Therefore, for each head movement a unique pattern of excitation and inhibition results, which is recognized by the brain and interpreted as the position of the head with respect to gravity.

Acknowledgements

Grateful acknowledgement is made to the following sources for permission to reproduce material in this book:

Cover

Copyright © Mehau Kulyk Science Photo Library.

Figures

Figure 2.5: Schiffman, R. (2000) Copyright © John Wiley and Sons Inc., *Sensation and Perception: An Integrated Approach*, 5th edn. Used by permission of John Wiley & Sons Inc; *Figure 2.5 (human voice)*: Kinsler, L. E. and Frey, A. R. (1962) Copyright © John Wiley and Sons, Inc. *Fundamentals of Acoustics*, 2nd edn. Used by permission of John Wiley & Sons, Inc; *Figure 3.4(a)*: Picture by Mireille Lavigne-Rabillard, from 'Promenade around the cochlea', by R. Pujol, S. Blatrix, T. Pujol and V. Reclar-Enjalbert, (www.iurc.montp.inserm.fr/cric/audition/index.htm) CRIC, University Montpellier; *Figure 3.7(b)*: Copyright © Science Photo Library; *Figure 3.10*: Bekesy, G. v. (1953) 'Description of some mechanical properties of the organ of corti', *The Journal of the Acoustical Society of America*, vol. 25, no. 4, July 1953, American Institute of Physics; *Figure 3.11(b)*: Zemlin, W. R. (1981) *Speech and Hearing Science: Anatomy and Physiology*, 2nd edn, Prentice-Hall, Inc. Copyright © 1981, 1968 by Prentice-Hall, Inc., Englewood Cliffs, N.J. 07632; *Figure 3.14*: Yost, W. A. (2000) 'Peripheral auditory nervous system and haircells', *Fundamentals of Hearing: An Introduction*, 4th edition, Academic Press. Copyright © 2000 by Academic Press; *Figure 3.19*: Kiang, N. Y. S. (1980) 'Processing of speech by the auditory nervous system', *The Journal of the Acoustical Society of America*, vol. 68, American Institute of Physics; *Figure 4.1*: Kandel, E. R., Schwartz, J. H. and Jessell, T. M. (2000) *Principles of Neural Science*, 4th edition, McGraw-Hill. Copyright © 2000 by The McGraw-Hills Companies, Inc. All rights reserved; *Figure 4.3(a)*: Rose, J. E., Hind, J. E., Anderson, D. J. and Brugge, J. F. (1971) 'Some effects of stimulas intensity on response of auditory nerve fibers in the Squirrel Monkey', *Journal of Neurophysiology*, **34**, The American Physiological Society; *Figure 4.3(b)*: Goldstein, E. B. (1999) *Sensation and Perception*, 5th edition, Brooks/Cole Publishing Company. Copyright © 1999 by Brooks/Cole Publishing Company. A division of International Thomson Publishing, Inc; *Figure 4.4*: Adapted from Wever, E. G. (1949) *Theory of Hearing*, John Wiley & Sons, Inc. Copyright © 1949 John Wiley & Sons, Inc. Reprinted by permission; *Figure 4.7*: Lindsay, P. H. (1972) *Human Information Process: An Introduction to Psychology*, Academic Press. Copyright © 1972 by Academic Press; *Figure 4.10*: Bear, M. F., Connors, B. W. and Paradiso, M. A. (1996) *Neuroscience: Exploring the Brain*, Williams & Wilkins. Copyright © 1996 Williams & Wilkins; *Figure 5.3*: Sivian, L. J. and White, S. D. (1933) 'On minimum audible sound fields', *The Journal of the Acoustical Society of America*, vol. 4, American Institute of Physics; *Figure 5.4*: Fletcher, H. and Munson, W. A. (1933) 'Loudness, its definition, measurement and calculation, *The Journal of the Acoustical Society of America*, vol. 5, October 1933, American Institute of Physics; *Figure 5.5*: Stevens, S. S., Volkmann, J. (1940) 'The relation of pitch to frequency: a revised scale', *The American Journal of Psychology*, vol. 53, no. 3, July 1940, The University of Illinois Press; *Figure 5.6*: Harris, J. D. (1952) 'Pitch discrimination', *The Journal of the*

Glossary for Block 3

absolute limen *See* absolute threshold.

absolute pitch *See* perfect pitch.

absolute refractory period The period of time, measured from the onset of an action potential, during which another action potential cannot be generated.

absolute threshold The minimum stimulus that evokes a response.

acoustic impedance A measure of how readily the particles of the conducting medium can be displaced by sound waves. It is measured as the ratio of the sound pressure to the flow velocity of the particles of the transmitting medium.

acoustic neuroma *See* vestibular schwannoma.

acoustic reflex *See* middle ear reflex.

acoustic shadow *See* sound shadow.

acoustic threshold The amplitude of a sound pressure wave that can just be heard when presented to a listener.

activation curve The relationship between the number of mechanosensitive channels open and hair bundle displacement.

acute otitis media Acute inflammation of the middle ear.

adaptation Change in sensitivity in response to continued stimulation. For hair cells, the shifting of the activation curve to a new working point as a result of maintained deflection of the hair bundle. This ensures that the sensitivity of hair cells is always at a maximum close to the resting position.

afferent (nerve) fibre A nerve fibre that carries neural information from the periphery to the brain.

ampliotopic organization The systematic gradient of response selectivity to sound level.

amplitude The maximum departure of an oscillating system from its average value.

ampulla The bulge at one end of each semicircular canal which contains the hair cells that are stimulated by rotation of the head.

anteroventral cochlear nucleus The anteroventral division of the cochlear nucleus.

apical Towards the apex.

audiogram An auditory threshold curve. A plot of the minimum stimulus that evokes a response for a range of frequencies.

auditory cortex The part of the brain that is involved with processing auditory information. Located in the temporal lobe adjacent to a deep groove or convolution in the brain called the lateral (sylvian) fissure.

auditory discrimination The ability to pick out a sound in a noisy environment and concentrate on that sound while ignoring the background noise. Can also be used to refer to the ability to discern a difference between two sounds.

auditory excitation pattern The pattern of basilar membrane movement in response to complex sounds; usually plotted as a graph on a frequency scale represented as distance along the basilar membrane against response level in dB.

auditory nerve (cochlear nerve) The part of the VIIIth cranial nerve that carries information from the auditory system to the brain (via afferent neurons) and from the brain to the auditory system (via efferent neurons).

auditory threshold *See* absolute threshold.

autocorrelation The technique used to compare a waveform with a delayed version of itself.

band-pass filter A wave filter that has a single transmission band extending from a lower cut-off frequency greater than zero to a finite upper cut-off frequency.

basal Towards the base.

basilar membrane A membrane that runs the length of the cochlear canal. It separates the scala media from the scala tympani and supports the organ of Corti.

beats The regular changes in amplitude of a sound that results when two (or more) sine waves of similar frequency are added together.

bipolar neurons Neurons with just two processes.

bone conduction The conduction of sound to the inner ear through the temporal bone.

brainstem vestibular nuclei *See* vestibular nucleus.

Broca's area The inferior prefrontal cortex of the brain, an area involved in linguistic analysis, specifically auditory word perception and repetition.

bushy cells Found in the ventral cochlear nuclei and so named because each has a single, stout, modestly-branched primary dendrite that is adorned with numerous fine branchlets. Electrical stimulation of a bushy cell characteristically elicits only one action potential. They provide information about the timing of acoustical stimuli.

calcium-activated potassium channels Special potassium channels in the hair cell which open in response to calcium ion influx thereby allowing potassium ions to leave the cell.

Central Auditory Processing Disorder (CAPD) The condition in which individuals perform standard hearing tests normally but have difficulty in discriminating sound in noisy environments, such as found in classrooms.

central frequency (CF) *See* characteristic frequency.

characteristic frequency (CF) The sound frequency to which a neuron in the auditory system gives the greatest response.

cholesteatoma A small bag of chronic inflammation that may progressively infect and damage the structures of the middle ear; caused when persistent negative pressure in the middle ear sucks part of the eardrum inwards and the inside-out portion of eardrum traps layers of dead skin which then become infected.

cochlea A spirally coiled, tapered bony tube of about two and three-quarter turns located within the inner ear. It contains the receptors essential for hearing.

cochlear amplifier Outer hair cells, including the motor protein in the outer hair cell membrane, responsible for amplifying the displacements of the basilar membrane in the cochlea.

cochlear duct An endolymph-filled duct following the spiral shape of the cochlea, and containing the organ of Corti. The cavity of the cochlear duct is called the scala media.

cochlear ganglion *See* spiral ganglion.

cochlear implantation The insertion of a device into the cochlea to induce sound perception. The device is a multi-channel electrode and complex speech processing algorithms are used to stimulate the auditory nerve directly.

cochlear nerve *See* auditory nerve.

cochlear nuclear complex The first stage of the central auditory pathway, which is located in the brainstem and consists of the dorsal cochlear nucleus and the two divisions of the ventral cochlear nucleus.

cochlear nucleus A nucleus in the medulla that receives afferents from the spiral ganglion in the cochlea.

cochlear partition The term used collectively to describe the partitions of the scala media, the most significant of which are the basilar membrane, tectorial membrane and organ of Corti.

coincidence detectors Binaural neurons in the superior olive that fire maximally when ipsilateral and contralateral inputs arrive simultaneously.

concha The deep central portion of the pinna. It has a diameter of 1 to 2 cm.

conductive pathway The pathway that comprises the pinna, the external auditory meatus, the tympanic membrane and the middle ear.

convergent inputs Inputs received by neurons which have come from various sources (e.g. from both utricle and saccule, or from both canals and otolithic organs). Information from convergent inputs can be rapidly integrated, with a reduced possibility of error from spontaneous or random activity in the organs.

Corti's arch A rigid inverted V-shaped structure that rests on the basilar membrane. The hair cells are found on either side of the arch.

crista The sensory structure within the ampulla of the semicircular canals. The cristae respond to changes in the rate of movement of the head.

critical frequency (CF) See characteristic frequency.

cross-correlation The comparison of two waveforms where one is delayed relative to the other by finding the compensatory delay time necessary to cancel out the original delay. The mechanism proposed to explain our remarkable sensitivity to changes in interaural time difference. The delays required to cancel the ITD are produced by electrical impulses having to travel different distances down an axon. The subsequent firing pattern of coincidence detector neurons provides a place code.

cupula A gelatinous body in the ampulla of the semicircular canals in which the hair cells of the crista are embedded.

cycle The repeat unit of a waveform.

decibel The decibel is one-tenth of a bel. It describes the ratio of one quantity to another on a logarithmic scale. For power or intensity it is defined by $dB = 10 \log (I_1 / I_2)$ for the two intensities I_1 and I_2. For pressure it is defined by $dB = 20 \log (P_1 / P_2)$ for the two pressures P_1 and P_2. Since intensity is proportional to $(\text{pressure})^2$, the two measures are equivalent.

delay lines A component that introduces a delay in the transmission of a signal. In the superior olivary complex, axons act as delay lines.

descending auditory pathway Descending projections from the auditory cortex to the medial geniculate nucleus, the inferior colliculi and the brain stem auditory nuclei.

difference limen (DL) The minimum change in stimulus that can be correctly judged as different from a reference stimulus in a specified fraction of trials.

difference threshold *See* difference limen.

dorsal cochlear nucleus The dorsal division of the cochlear nucleus.

dynamic range The maximum response range to some physical variable. In hearing, it refers to the range of frequencies or intensities to which the human auditory system is sensitive.

efferent (nerve) fibre A nerve fibre that carries neural information from the brain to the periphery.

electrophysiology The various techniques used to measure the electrical changes caused by the activity of neurons in the central and peripheral nervous system. An example is electroencephalography, which involves measurement of the currents in the brain using electrodes attached to the scalp.

end-bulb of Held The final terminal made by each Type I cochlear afferent neuron in the anteroventral cochlear nucleus. They ensure rapid and precise transmission of every incoming action potential to the spherical bushy cells.

endocochlear potential The potential of about 80 mV that exists between the endolymph surrounding the stereocilia and the perilymph bathing the basal region of the hair cell.

endolymph A watery fluid contained within the membranous labyrinth. It is thought to be secreted by the stria vascularis.

envelope The overall amplitude change in a waveform. The envelope is formed by drawing one line connecting the peaks of the waveform and another connecting the troughs.

equal loudness contours Equal loudness contours define the phon scale. They provide the means of studying the relationship between loudness (a subjective quality) and sound pressure level, which is a physical quantity.

Eustachian tube The small tube that connects the middle ear to the back of the throat (the pharynx). It equalizes the pressure on each side of the tympanic membrane.

excitotoxic The toxicity that excessive amounts of a substance presents to cells. Excessive amounts of glutamate are highly toxic to neurons, known as glutamate excitotoxicity.

external auditory canal (meatus) The canal that conducts sound vibrations from the pinna to the tympanic membrane.

external ear An alternative name for the outer ear.

exteroception The collection and processing of signals originating outside the body (vision, audition, touch, smell and taste).

extraocular muscles Three pairs of muscles attached to the eye which control the rotation of the eye in three planes.

filter A component that removes a discrete range of frequencies from a wave source, and allows the remainder to pass unchanged.

footplate A flat oval bone that forms part of the stapes. The footplate is implanted in the oval window.

formant frequencies In humans, the vocal tract (between the larynx and the lips) resonates at different frequencies so that different parts of the frequency spectrum are amplified. These amplified frequencies are known as formant frequencies. Musical instruments also have formant frequencies, which stay constant regardless of the note being played.

Fourier analysis The process of decomposing a waveform into its sinusoidal components.

Fourier series The mathematical series (also called a harmonic series) obtained by carrying out a Fourier analysis on a complex periodic waveform. For example, a periodic waveform made up of harmonics of 100 Hz will have a Fourier series specified by the amplitude and phase of sinusoidal wave components at 100 Hz, 200 Hz, 300 Hz, etc. If the waveform is not periodic then the result of a Fourier analysis is a Fourier transform, a continuous function not a series.

Fourier transform The characterization of an aperiodic signal by its frequency content. An aperiodic signal is one that does not go through repetitive cycles.

frequency code Neurons in the auditory system fire action potentials at intervals that reflect the frequency of the input stimulus.

frequency spectrum The distribution with respect to frequency of the magnitudes of the components of a sound wave. It can be represented by plotting power, amplitude or level as a function of frequency.

frequency The number of cycles of a periodic process occurring per unit time. The SI unit of frequency is the hertz (Hz), where $1\ \text{Hz} = 1\ \text{s}^{-1}$.

functional magnetic resonance imaging (fMRI) This is an indirect functional imaging technique which, in one version, is able to detect local increases in blood flow in activated areas of the brain and display these on a conventional MRI scan.

fundamental frequency The lowest frequency component in a harmonic series.

fusiform cells Found in the dorsal cochlear nucleus and thought to participate in the localization of sound sources along the vertical axis.

glutamate An amino acid that acts as a neurotransmitter. In the ear, glutamate is released by hair cells at the afferent synapse as a result of depolarization of the hair cell, thereby initiating an action potential in the associated neuron.

glutamate transporter molecules Proteins which take up extracellular glutamate, terminating its action and making it ready for further activation. The take-up of glutamate prevents it becoming excitotoxic (toxic to neurons).

grommet A tiny plastic or metal ventilation tube inserted into the tympanic membrane to allow air to get from the atmosphere into the middle ear cavity.

hair bundles *See* stereocilia.

hair cells Sensory receptor cells for hearing and balance. Ciliated epithelial cells located within the organ of Corti (hearing) and the vestibular system (balance). In the organ of Corti there are two types: inner and outer hair cells. The bases of the hair cells are in contact with the dendritic processes of the neurons of the spiral ganglion of the cochlea.

harmonic A sinusoidal waveform having a frequency that is an integral multiple of the fundamental frequency of the periodic signal to which it is related.

harmonic frequencies A frequency that is a whole number multiple of the fundamental frequency.

harmonic series *See* Fourier series.

harmonically related Two waves are harmonically related if their frequencies are both whole number multiples of some fundamental frequency.

hearing threshold The quietest sound that an individual can hear in a soundproofed environment.

helicotrema A narrow aperture within the apex of the cochlea that allows communication between the scala vestibuli and the scala tympani.

hertz (symbol **Hz**) The SI unit of frequency: one hertz is one cycle every second.

Heschl's gyrus An area of the auditory cortex running transversely across the supratemporal plane; the primary auditory cortex is located here and virtually all simple and complex sounds produce activation of this area.

high-pass filter A wave filter having a single transmission band extending from some critical or cut-off frequency, not zero, up to infinite frequency.

incus The central member of the three auditory ossicles that are located in the middle ear. The body of the incus is attached to the head of the malleus and the rounded projection at the lower end of the incus is attached to the head of the stapes.

inferior cerebellar peduncle A large fibre tract connecting the cerebellum to the brain stem.

inferior colliculus A nucleus in the midbrain from which all ascending auditory signals project to the medial geniculate nucleus. The inferior colliculus is involved with hearing and pairs up with the superior colliculus, which is involved with vision.

inferior nucleus One of the four vestibular nuclei, it is located in the rostral medulla and contains small to medium-sized neurons.

infrasonic Sounds with frequencies below 20 Hz.

inner ear Part of the ear that consists of the cochlea and vestibular system.

inner hair cells *See* hair cells.

intensity The sound power transmitted through a given area in a sound field. It is expressed in watts per square metre. The term is also used as a generic name for any quantity relating to the amount of sound, such as power or energy, although this is not technically correct.

intensity level The intensity of a sound (in decibels) is 10 times the logarithm (to the base 10) of the ratio of the intensity of this sound to the reference intensity. The reference intensity commonly used in acoustics is 10^{-12} W m^{-2} (the threshold level for an average young adult, at 1 kHz).

interaural intensity difference *See* interaural level difference.

interaural level difference (ILD) The difference in intensity of sound at the two ears.

interaural time delay *See* interaural time difference.

interaural time difference (ITD) The time difference between a sound reaching one ear and then the other ear.

ion channel A pore formed by a membrane-spanning protein which allows the passage of ions from one side of the membrane to the other.

isofrequency layers Layers in the central nucleus of the inferior colliculus, made up of bands of cells with flattened dendrite fields in which the neurons are tuned to the same frequency. High-frequency layers are found towards the mid-line, low-frequency layers towards the outside.

iso-intensity contour The curve that results from recording the firing rate of a neuron in response to a range of tones of different frequency presented at equal sound pressure level.

just noticeable difference (JND) The smallest detectable difference between two stimuli (also called the JND or difference limen).

lateral fissure The pronounced groove on the surface of the brain that separates the frontal lobe from the temporal lobe. Also called the sylvian fissure.

lateral lemniscus A prominent neural tract that carries axons (a lemniscus is a collection of axons) of the superior olivary nucleus to the inferior colliculus of the midbrain.

lateral nucleus One of the four vestibular nuclei, it is located in the medulla and contains mainly large multipolar neurons.

longitudinal wave Displacement of the medium through which the wave travels is in the direction of propagation of the wave (as opposed to perpendicular to the direction of the propagation of the wave when it is known as a transverse wave). Sound waves travel as successive regions of compression and rarefaction of the air or another medium. They are therefore called longitudinal waves because the transmitting molecules oscillate back and forth in the same direction as the wave's propagation.

low-pass filter A wave filter having a single transmission band extending from zero up to some critical or cut-off frequency.

macula A sensory epithelium found in the otolithic organs (saccule and utricle) of the inner ear.

magnetoencephalography (MEG) This is a direct functional imaging technique that is able to detect, and locate, the very weak magnetic fields associated with the electrical currents caused by neuronal activity.

malleus The outermost of the three auditory ossicles that are located in the middle ear. The handle of the malleus is attached to the tympanic membrane and the head is attached to the body of the incus.

masking The amount by which the threshold of audibility of a sound is raised by the presence of another (masking) sound.

meatus *See* external auditory canal.

mechanical spectrum analyser The function of the basilar membrane. The mechanics of the basilar membrane are such that different frequencies excite hair cells in different areas of the membrane, with high frequencies at the base and low frequencies at the apex.

mechano-electrical transduction The conversion of a mechanical stimulus (movement of the stereocilia) into an electrical signal (the receptor potential).

mechanosensitive ion channels Channels in the hair cells which are activated by movement of the stereocilia, allowing entry of positively-charged ions which depolarize the cells.

medial geniculate body (nucleus) A relay nucleus in the thalamus through which all auditory information passes on its way from the inferior colliculus to the auditory cortex.

medial longitutidinal fasciculus A complex of connections among the nuclei of the cranial nerves that control movement of the eyes (the oculomotor, trochlear and abducens nerves).

medial nucleus One of the four vestibular nuclei, it is located in the rostral medulla and contains small to medium-sized neurons.

mel A unit of pitch. One thousand mels is the pitch of a 1000 Hz pure tone for which the loudness level is 40 phons. The pitch of a sound that is judged by a subject to be *n* times that of a 1000 mel tone is *n* thousand mels. Thus, the pitch of a sound subjectively judged to be 2 times that of a 1000 mel tone is 2000 mels.

membranous labyrinth A membranous compartment inside the bony labyrinth of the inner ear, surrounded by an outer membrane sheath.

Menières disease A disorder of the inner ear characterized by episodes of hearing loss, tinnitus and vertigo.

method of constant stimuli A psychophysical method used primarily to determine thresholds. In this procedure a number of stimuli ranging from rarely, to almost always perceivable, are presented one at a time. The subject responds to each presentation 'yes' or 'no'.

method of limits A psychophysical method used primarily to determine thresholds. In this procedure some dimension of the stimulus, or of the difference between two stimuli, is varied incrementally until the subject changes his/her response.

middle ear The air-filled chamber within the mastoid portion of the temporal bone, which contains the three auditory ossicles.

middle ear ossicles The three small bones found in the middle ear.

middle ear reflex The spontaneous contraction of the muscles in the middle ear in response to the sudden onset of a loud sound.

modiolus The conically-shaped central core of the cochlea. It contains the spiral ganglion of the cochlea and forms the inner wall of the scala vestibuli and scala tympani.

motor proteins These are found in the membrane of the outer hair cells. They amplify the movement of the basilar membrane in response to low-intensity sounds by changing the length of the cells. The nature of motor proteins is unknown.

multipolar cells *See* stellate cells.

notch filter A wave filter that blocks one frequency band and passes both higher and lower frequencies.

Obscure Auditory Dysfunction (OAD) *See* Central Auditory Processing Disorder (CAPD).

octopus cells Neurons with thick dendrites and axons which are found in the ventral cochlear nucleus, and which only respond to the onset of a signal.

oculomotor nuclei Nuclei found in the brainstem, which are involved in the vestibulo-ocular reflex.

oculomotor system The system that controls accurate and coordinated movements of the eye by the extraocular muscles in order to direct attention toward a particular visual stimulus.

olivocochlear bundle Efferent nerve fibres carried by the cochlear nerve from the brain stem to the cochlea. The fibres come from nerve cells around the outside of the superior olivary complex.

organ of Corti The auditory receptor organ, which contains a series of neuro-epithelial hair cells (receptor cells for hearing) and their supporting structures lying on top of the basilar membrane and beneath the tectorial membrane. It is located within the scala media of the cochlea and extends from the base of the cochlea to the apex.

ossicles *See* middle ear ossicles.

otitis media (with effusion) A condition arising when the Eustachian tube malfunctions and causes the middle ear to fill up with fluid that is produced by the lining of the cavity itself. (If the fluid is thick and viscid, the condition is given the colloquial name 'glue ear'.) Otitis media with effusion is the most common cause of conductive hearing loss in the UK.

otoacoustic emissions Small amounts of sound emitted by the ears in response to sound input and used as the basis of a hearing test for neonates.

otolithic membrane A delicate acellular structure covering the macula.

otolithic organs The saccule and the utricle of the inner ear which detect the force of gravity and linear acceleration.

otoliths Calcium carbonate crystals which cover the surface of the otolithic membrane.

otosclerosis A condition resulting from the formation of new bone around the footplate of the stapes that fuses the bone rigidly with the bone of the inner ear. This reduces the ear's ability to conduct sound from the ossicular chain to the sensory receptors of the cochlea.

outer ear Consists of the pinna, the external auditory canal (meatus), and the tympanum (eardrum).

outer hair cells *See* hair cells.

outer spiral fibres Dendrites of Type II ganglion cells which contact the outer hair cells. Each outer spiral fibre contacts up to ten outer hair cells.

oval window An opening through the bone that separates the middle ear from the scala vestibuli of the cochlea. It is closed by a membrane to which is connected the footplate of the stapes.

parietal cortex The outer cortical layer of the parietal lobes; involved in specialized spatial processing and activated by sounds containing spatial and motion cues.

perfect pitch (absolute pitch) The ability to recognize the pitch of a note in isolation, without an external reference.

perilymph The fluid that fills the scala vestibuli and scala tympani in the cochlea in the inner ear, containing low K^+ and high Na^+ concentrations.

period The time taken to complete one cycle of a quantity that varies periodically with time.

periodic waveform A waveform that continuously traces the same path. Periodic signals continue to repeat indefinitely.

peripheral auditory system The sound processing components consisting of the outer, middle and inner ear, and the auditory nerve.

phase Relative position (in time or space) of two or more sinusoidal waves. There are 360 degrees in one full cycle of a sine wave. Two sinusoidal waves are said to be in phase when they start at the same point in their cycle, and the phase difference between them is zero degrees or an integral multiple of 360 degrees.

phase ambiguity Caused when the wavelength of a sound is shorter than the distance between the ears (i.e. high frequencies). Differences in the time at which the same phase of a sound wave reaches each ear can no longer be resolved by the auditory system.

phase locking The consistent firing of an auditory neuron at the same phase of a sound wave.

phon A unit of loudness. (*See also* equal loudness contours.)

pinna The pinna is the most visible part of the ear. It is a fibrocartilaginous plate that is attached to the head and is useful in the localization of sounds in both the front–back and in the vertical dimensions.

pitch The psychological attribute of sound most closely associated with the frequency of the sound.

place code The idea that the frequency of a tone is signalled by the place in the auditory system that is maximally stimulated.

place principle *See* place code.

place theory Theory that different positions along the basilar membrane are sensitive to different frequencies.

planum temporale An area of the auditory cortex, lateral to Heschl's gyrus, involved in movement processing.

positron emission tomography (PET) An indirect functional imaging technique in which functionally-induced changes in brain metabolism are monitored by the preferential take up of a radioactive isotope that decays by positron emission.

posteroventral cochlear nucleus The posteroventral division of the cochlear nucleus.

presbyacusis The hearing loss associated with a decrease in cochlear sensitivity with age.

primary auditory cortex (A1) Located in the lateral fissure, along the upper surface of the temporal lobe, the primary auditory cortex is surrounded by the secondary auditory cortex, which is subdivided into six distinct areas.

primitive grouping mechanisms The way in which the listener uses generally useful unlearned cues to separate sound into different sources.

proprioception The sensation of body position and movement using sensory signals from muscles, joints and skin.

psychometric function A mathematical relationship in which the independent variable is a measure of stimulus and the dependent variable is a measure of response.

pure tone A tone produced by a periodic vibration at a single frequency. Pure tones are approximated by good tuning forks.

radial fibres Dendrites of Type I ganglion cells, which contact the inner hair cells. Each inner hair cell synapses with up to twenty radial fibres.

receptor potential A stimulus-induced change in the membrane potential of a sensory receptor.

reference intensity Required when using the decibel scale to measure the intensity of a sound. The reference intensity most commonly used in acoustics is $10^{-12}\,\mathrm{W\,m^{-2}}$.

reference pressure Required when using the decibel scale to measure the amplitude of a sound. The reference pressure most commonly used when working with sound is 20 μPa. When this reference is used the level is known as sound pressure level (SPL).

Reissner's membrane A delicate membrane that separates the scala media from the scala vestibuli and forms one wall of the cochlear duct.

resonant frequency The value of the frequency of an oscillation that creates the condition of resonance in which the amplitude is a maximum.

retrocochlear Relating to the neural pathway beyond the cochlea, the auditory nerve.

round window An opening through the bone that separates the middle ear from the scala tympani of the cochlea. It is located behind and below the oval window and is closed by the round window membrane.

saccule One of the two otolithic organs located in the inner ear; it detects changes in head angle and linear acceleration.

saturated A neuron is said to be saturated when it no longer increases its firing rate in response to an increase in stimulus intensity.

scala media The cavity of the cochlear duct filled with endolymph and also containing the tectorial membrane. The scala media is separated from the scala vestibuli by Reissner's membrane and from the scala tympani by the basilar membrane. The scala media has a closed end at the cochlea apex.

scala tympani The perilymph-filled passage of the cochlear canal that extends from the round window at the base to the helicotrema at the apex. It is separated from the scala media by the basilar membrane and the cellular structures attached to it.

scala vestibuli The perilymph-filled passage of the cochlear canal that extends from the oval window at the base to the helicotrema at the apex.

Scarpa's ganglion Also known as the vestibular ganglion. It consists of the cell bodies of the vestibular neurons which leave the hair cells of the vestibular labyrinth and travel to the vestibular nuclei in the brainstem.

schema-based mechanisms The way in which the listener uses detailed knowledge about what specific sounds are like to select a relevant subset of the total sound.

secretory otitis media *See* otitis media (with effusion).

semicircular canals Part of the inner ear. They detect turning movements of the head (angular acceleration).

signal detection The discrimination of one signal of interest from a number of other signals. It is believed that the descending auditory pathway plays a role in this.

sine-wave speech A transformation of natural speech that consists of three (frequency-modulated) sine waves on which the speech frequencies are superimposed.

sound pressure level (SPL) The ratio (in decibels) of the pressure of a sound wave to the reference pressure of 20 μPa. 20 μPa is the average minimum pressure the human ear can detect. The corresponding value for intensity is the ratio (in decibels) of the intensity of a sound wave to the reference intensity of $10^{-12}\,\mathrm{W\,m^{-2}}$.

sound shadow A region in a field in which the sound is reduced owing to an object which blocks, obscures, absorbs or reflects the physical energy. The sound shadow produced by one's head serves as a cue in the location of sounds of relatively high frequencies.

spherical bushy cells *See* bushy cells.

spiral (cochlear) ganglion Located within the modiolus, it is composed of cell bodies of the neurons of the cochlear nerve. The dendritic processes of these bipolar neurons make synaptic contact with the hair cells. The axons terminate in the cochlear nucleus in the medulla.

stapedius muscle The intra-aural muscle attached to the neck of the stapes. Its reflex response to an intense sound stimulus is to swing the footplate of the stapes outwards and backwards from the oval window. The joint action of the stapedius muscle and the tensor tympani muscle is to limit the motion of the auditory ossicles and thereby help protect the inner ear from damage by intense sound, especially at low frequencies.

stapes The innermost and smallest of the three auditory ossicles which are located in the middle ear. Its shape resembles a stirrup. The head of the stapes is attached to the incus, and the footplate nearly fills the oval window to which it is attached by the annular ligament.

stellate cells Found in the ventral cochlear nuclei, stellate cells have many symmetrical dendrites and respond to a stimulus with a train of regularly-spaced action potentials. They encode frequencies present in a given auditory input.

stereocilia A bundle of hair-like processes attached to the top of the hair cells of the inner ear. Each stereocilium is a rigid cylinder consisting of actin filaments cross-linked to the protein fibrin.

stimulus frequency principle The most distinctive organizing principle in the auditory pathway. The component frequencies associated with each auditory stimulus stimulate specific areas of the basilar membrane. Each auditory nucleus has a tonotopic organization, replicating the progressive change in frequency that occurs along the length of the basilar membrane.

stria vascularis An epithelium located on the lateral wall of the cochlear duct which transports endolymph to the scala media.

striola A curved zone in the maculae of the utricle and saccule which contains a large proportion of small hair bundles. It delineates the reversal in orientation of the hair cells of the maculae.

summating potential The sustained depolarization of hair cells caused by a sound stimulus at frequencies above 1000 Hz.

superior colliculi The front pair of four bumps located on the dorsal surface of the midbrain. The superior colliculi are involved with vision; the rear pair, the inferior colliculi, are involved with hearing.

superior nucleus One of the four vestibular nuclei, it is located in the caudal pons and contains small to medium-sized neurons.

superior olivary complex The first place in the ascending auditory pathway where information from both ears is combined. It is mainly involved with sound localization.

superior temporal gyrus The area within the temporal lobe containing the auditory cortex. This region is one of the most highly folded areas of the human brain; gyrus (pl. gyri) is the name given to the crests.

superior temporal sulcus The area within the temporal lobe associated with higher-level processing of acoustic information. This region is one of the most highly folded areas of the human brain; sulcus (pl. sulci) is the name given to the valleys.

supratemporal plane The upper surface of the superior temporal gyrus.

sylvian fissure A deep fold constituting the uppermost boundary of the temporal lobe. Also called the lateral fissure.

tectorial membrane A soft semi-gelatinous ribbon-like structure attached along one edge to the spiral limbus and along the other edge to the outer border of the organ of Corti. It is in intimate contact with the cilia of the hair cells.

temporal lobe Located on the side of the brain, above the ears. It functions in hearing, memory, vision and the categorization of objects.

tensor tympani muscle The intra-aural muscle attached to the handle of the malleus. Its reflex contraction, in response to intense sound or tactile stimulation to parts of the face, draws the malleus inwards, which increases the tension on the tympanic membrane. The joint action of the tensor tympani muscle and the stapedius muscle is to limit the motion of the auditory ossicles and thereby help protect the inner ear from damage by intense sound, especially at low frequencies.

threshold The smallest value of a stimulus that can be detected. A measure of sensitivity.

timbre A characteristic quality of sounds produced by a particular voice or instrument that depends on the number and quality of the overtones; partly determined by the relative amplitudes of the different harmonics.

tinnitus A condition often colloquially described as 'a ringing in the ears'. Formally acknowledged to be the perception of short-lived or persistent episodes of sound that originate in the head; it may represent phantom auditory perception but effective treatments remain elusive.

tip link A filamentous connection between two hair cell stereocilia.

tonotopic mapping The coding of sound frequency to position of excitation.

tonotopic organization A systematic organization within an auditory structure on the basis of characteristic frequency.

transformer action The mechanism by which sound waves travelling in air in the outer ear are converted to sound waves travelling in fluid in the inner ear without being reflected at the interface. The transformer action is provided by a combination of the difference in area between the tympanic membrane and the stapes footplate and the lever action of the ossicles in the middle ear.

transverse wave Displacement of the medium through which the wave travels is perpendicular to the direction of the propagation of the wave. Waves on water and electromagnetic waves are transverse.

travelling wave A wave that propagates energy is known as a travelling wave (as opposed to a standing or stationary wave which does not).

tuning curve A curve that describes how the threshold intensity (the intensity required for the neuron to fire an action potential) for a given fibre (or hair cell) varies as a function of stimulus frequency.

tympanic membrane Also known as the eardrum or tympanum. A conically-shaped semi-transparent membrane that separates the external auditory canal (meatus) from the middle ear cavity. The handle of the malleus is attached to it in the middle ear.

tympanosclerosis Scarring of the eardrum as a result of infection.

tympanum *See* tympanic membrane.

ultrasonic Sounds with frequencies above 20 000 Hz.

utricle One of two otolithic organs found in the inner ear; it detects changes in head angle and linear acceleration.

vestibular apparatus *See* vestibular system.

vestibular nerve Part of the VIIIth cranial nerve that carries information from the vestibular system to the brain (via afferent neurons) and from the brain to the vestibular system (via efferent neurons).

vestibular nucleus A nucleus in the medulla that receives input from the vestibular labyrinth of the inner ear.

vestibular schwannoma Benign tumours arising on the vestibular nerve and derived from the Schwann cells that surround and insulate the nerve fibres. They usually cause unilateral hearing loss and tinnitus.

vestibular system A part of the inner ear specialized for the detection of head motion. It consists of the otolithic organs and the semicircular canals.

vestibule A perilymph-filled oval-shaped central cavity of the osseous labyrinth. It contains the oval window, which is located in its wall facing the middle ear cavity (tympanic wall), as well as the utricle and the saccule.

vestibulocochlear nerve The VIIIth cranial nerve that carries information from the inner ear (cochlea and vestibular system) to the brain (via afferent neurons) and from the brain to the inner ear (via efferent neurons).

vestibulocollic reflex (VCR) A reflex movement of the head which arises as a result of information from the vestibular system, and serves to stabilize the head against forces which would move it.

vestibulocollic reflex pathway The pathway through which the VCR is activated.

vestibulo-ocular reflex (VOR) A reflex movement of the eyes which is stimulated by rotational movements of the head. It stabilizes the visual image on the retina.

vestibulo-ocular reflex pathway The pathway through which the VOR is activated.

vestibulospinal tract Part of the medial brainstem pathway that runs from the vestibular nucleus to the spinal cord, and is involved in the control of posture in response to information from the vestibular system.

voice onset time A temporal feature that helps to distinguish between two classes of stop consonants, voiced (e.g. 'ba'), and voiceless (e.g. 'pa').

volley principle The idea that high sound frequencies are represented in the pooled activity of a number of neurons, each of which fires in an intermittent though phase-locked manner.

wavelength The distance between successive peaks (or troughs) of a wave when plotted with distance along the horizontal axis.

Weber's law Weber's law states that the JND is a constant fraction of the standard stimulus: $\Delta I/S = K$, where K is the Weber fraction.

Wernicke's area The temporoparietal cortex, part of the network of brain areas concerned with linguistic analysis, particularly the analysis of lexical, semantic and syntactic information.

white noise A sound that contains a very large (infinite) range of frequencies.

Index

Entries and page numbers in **bold type** refer to key words which are printed in **bold** in the text and which are defined in the Glossary.